Chitin and Chitosan - Isolation, Properties, and Applications

Edited by Brajesh Kumar

Published in London, United Kingdom

Chitin and Chitosan - Isolation, Properties, and Applications
http://dx.doi.org/10.5772/intechopen.100948
Edited by Brajesh Kumar

Contributors
Renuka Vinothkumar, Janet Paterson, Mohsin O. Mohammed, Nadia Haj, Luz Irene Rojas-Avelizapa, Paul Edgardo Regalado-Infante, Norma Gabriela Rojas-Avelizapa, Daniel Tapia-Maruri, Andrea Margarita Rivas-Castillo, Rosalia Nuñez-Pastrana, Régulo Carlos Llarena-Hernández, Magdi A. E. Abdellatef, Eman Elagamey, Said M. Kamel, Pranab Dutta, Madhusmita Mahanta, Arti Kumari, Márcia Aparecida Sperança, Felipe Trovalim Jordão, Aline Diniz Cabral, Felipe Baena Garcia, Edmar Silva Santos, Rodrigo Buzinaro Suzuki, Max Mario Fuhlendorf

First published in London, United Kingdom, 2023 by IntechOpen
IntechOpen is the global imprint of INTECHOPEN LIMITED, registered in England and Wales, registration number: 11086078, 5 Princes Gate Court, London, SW7 2QJ, United Kingdom
Printed in Croatia

British Library Cataloguing-in-Publication Data
A catalogue record for this book is available from the British Library

Additional hard and PDF copies can be obtained from orders@intechopen.com

Chitin and Chitosan - Isolation, Properties, and Applications
Edited by Brajesh Kumar
p. cm.

This title is part of the Biochemistry Book Series, Volume 45
Topic: Chemical Biology
Series Editor: Miroslav Blumenberg
Topic Editors: Şükrü Beydemir and Deniz Ekinci

Print ISBN 978-1-80356-692-4
Online ISBN 978-1-80356-693-1
eBook (PDF) ISBN 978-1-80356-694-8
ISSN 2632-0983

IntechOpen Book Series
Biochemistry
Volume 45

Aims and Scope of the Series

Biochemistry, the study of chemical transformations occurring within living organisms, impacts all of the life sciences, from molecular crystallography and genetics, to ecology, medicine and population biology. Biochemistry studies macromolecules - proteins, nucleic acids, carbohydrates and lipids –their building blocks, structures, functions and interactions. Much of biochemistry is devoted to enzymes, proteins that catalyze chemical reactions, enzyme structures, mechanisms of action and their roles within cells. Biochemistry also studies small signaling molecules, coenzymes, inhibitors, vitamins and hormones, which play roles in the life process. Biochemical experimentation, besides coopting the methods of classical chemistry, e.g., chromatography, adopted new techniques, e.g., X-ray diffraction, electron microscopy, NMR, radioisotopes, and developed sophisticated microbial genetic tools, e.g., auxotroph mutants and their revertants, fermentation, etc. More recently, biochemistry embraced the 'big data' omics systems. Initial biochemical studies have been exclusively analytic: dissecting, purifying and examining individual components of a biological system; in exemplary words of Efraim Racker, (1913 –1991) "Don't waste clean thinking on dirty enzymes." Today, however, biochemistry is becoming more agglomerative and comprehensive, setting out to integrate and describe fully a particular biological system. The 'big data' metabolomics can define the complement of small molecules, e.g., in a soil or biofilm sample; proteomics can distinguish all the proteins comprising e.g., serum; metagenomics can identify all the genes in a complex environment e.g., the bovine rumen.
This Biochemistry Series will address both the current research on biomolecules, and the emerging trends with great promise.

Meet the Series Editor

Miroslav Blumenberg, Ph.D., was born in Subotica and received his BSc in Belgrade, Yugoslavia. He completed his Ph.D. at MIT in Organic Chemistry; he followed up his Ph.D. with two postdoctoral study periods at Stanford University. Since 1983, he has been a faculty member of the RO Perelman Department of Dermatology, NYU School of Medicine, where he is codirector of a training grant in cutaneous biology. Dr. Blumenberg's research is focused on the epidermis, expression of keratin genes, transcription profiling, keratinocyte differentiation, inflammatory diseases and cancers, and most recently the effects of the microbiome on the skin. He has published more than 100 peer-reviewed research articles and graduated numerous Ph.D. and postdoctoral students.

Meet the Volume Editor

Dr. Brajesh Kumar is currently an assistant professor and head of the Department of Chemistry, TATA College, Kolhan University, Chaibasa, India. He received a Ph.D. in Chemistry from the University of Delhi, India, in 2009. He is a pioneering researcher in the fields of green chemistry, nano synthesis, environmental remediation, photocatalysis, nanomedicine, and polymer and natural product extraction. Dr. Kumar has several national and international fellowships to his credit and has worked as a faculty member in various universities throughout India, Ecuador, and South Korea. He is a holder of two registered patents as well as a member of the American Chemical Society and the Indian Society of Chemists and Biologists (ISCB). He has also published more than eighty research articles and is an active reviewer of more than seventy journals. He was listed among the top 2% of scientists worldwide by Stanford University, USA, in 2021 and 2022.

Contents

Preface

Currently, numerous biomaterials-based studies are being conducted, including research into chitin and chitosan, the second most abundant polysaccharide after cellulose. Chitin is obtained at an industrial scale from a variety of natural sources, including crustacean and insect exoskeletons, fungi cell walls, squid pens, and others. Chitosan is biodegradable, biocompatible, non-toxic, and water-soluble under acidic conditions and linear cationic aminopolysaccharide derived from the deacetylation of chitin. It contains free amino and hydroxyl groups that can be functionalized by binding with the cationic and anionic groups. It has numerous applications, especially in environmental remediation and the biomedical, pharmaceutical, agriculture, and food industries. This book addresses isolation, properties, and certain applications of chitin and chitosan, including films, fibers, nanoparticles, composite materials, hydrogels, polymeric complexes, water purification, antimicrobials, textile, cosmetics, biosensors, nanoporous scaffolds, and membranes. It is written by researchers from industry, academia, government, and private research institutions.

This book highlights the isolation, properties, and applications of chitin and chitosan. In Chapter 1, Paul Edgardo Regalado-Infante et al. cover recent studies on the biological functions of chito-oligosaccharides and their impact on a priority area such as agriculture, where these compounds could be used to substitute for chemical compounds that have generated serious health issues as well as environmental pollution. In Chapter 2, Renuka Vinothkumar and Janet Paterson focus on a new way of developing water-soluble colloidal chitin (WSCC) from prawn waste and investigate its fundamental rheological and antibacterial properties. WSCC films studied during this research may be used in food packaging or in medical applications. The use of WSCC biodegradable films will protect the environment in the future and will be an effective alternative to plastics. In Chapter 3, Mohsin Mohammed and Nadia Haj loaded chitosan with 5-fluorouracil via amide-mediated binding and evaluated the resulting product as a potential 5-fluorouracil delivery agent. The authors use 1H-13C nuclear magnetic resonance and Fourier-transform infrared and ultraviolet spectroscopy to characterize the product. The three cancer cell lines MCF-7, MDA-MB-231, and MDA-MB-453 showed a dose-dependent reduction in viability compared to the original medicine and these findings suggest that chitosan–methotrexate as a prodrug could be helpful in the treatment of colon cancer.

Nanoscience will continue to grow due to its numerous benefits, including those on human health, food processing, environmental safety, and device engineering. In recent years, several findings have been gathered indicating nano-chitosan as a potential plant health material. In Chapter 4, Pranab Dutta, Arti Kumari, and Madhusmita Mahanta outline the methods of green synthesis and characterization of chitosan nanoparticles, their utilization in plant protection and growth promotion, and the underlying mechanisms. In the coming years, the use of nano-chitosan for combating biotic and abiotic stresses and transport of agrochemicals will be a promising discipline for utility in sustainable agriculture. In Chapter 5, Magdi A.E. Abdellatef et al. highlight that chitosan is the ideal resource for plant disease management under sustainable agriculture. Chitosan is biodegradable, biocompatible, non-toxic, and has antimicrobial activity, which confers it with dual synergetic effects during host–pathogen interaction. Leishmaniasis, an infectious disease that affects humans, domesticated dogs, and wild animals, is caused by 20 of the 53 Leishmania genus species and is transmitted by sandflies. Finally, in Chapter 6, Felipe Trovalim Jordão et al. investigate the potential of the chitinase-encoding gene as a molecular diagnostic tool and a phylogenetic marker for studying basal trypanosomatid groups worldwide.

I am grateful to my loving wife Kumari Smita for her helpful comments on several chapters and excellent support in the conceptualization of the book's content.

Dr. Brajesh Kumar
Department of Chemistry,
TATA College,
Kolhan University,
Jharkhand, India

Centro de Nanociencia y Nanotecnologia,
Universidad de las Fuerzas Armadas -ESPE,
Sangolqui, Ecuador

Chapter 1

New Perspectives on the Application of Chito-Oligosaccharides Derived from Chitin and Chitosan: A Review

Paul Edgardo Regalado-Infante,
Norma Gabriela Rojas-Avelizapa, Rosalía Núñez-Pastrana,
Daniel Tapia-Maruri, Andrea Margarita Rivas-Castillo,
Régulo Carlos Llarena-Hernández
and Luz Irene Rojas-Avelizapa

Abstract

The study of chitin and chitosan has stood out for many years due to their potential application in various areas such as the food industry, where they are either used as additives, prebiotics, or bio-conservatives; as to biomedical and pharmaceutical industries, where they function to treat diseases. Besides, in the agriculture field, it is known that they can cause a positive effect on the development of plants and optimize nitrogen fixation. In recent years, attention has been paid to their derivatives, chito-oligosaccharides which, unlike chitin and chitosan, they have different chemical characteristics, like their solubility, a characteristic that facilitates their use, contrary to chitin and chitosan. Moreover, the small size of chito-oligosaccharides can facilitate their entry into the cell. This review covers recent studies on the biological functions of chito-oligosaccharides and their impact on a priority area such as agriculture, where these compounds could be used to substitute the demand for chemical compounds that, until now, have generated serious health issues as well as environmental pollution.

Keywords: natural products, shrimp wastes, sustainable agriculture, chito-oligosaccharides, chitin and chitosan

1. Introduction

A world-class level concerning aspect is the accelerated population growth as well as an increased demand for goods and services. Against this background, it is important to define strategies that allow the supply of enough food for a population in continuous growth, through the implementation of efficient agricultural systems. Agriculture plays a key role in the development of any society; however, the various agrochemicals employed, such as biocides, growth stimulants, and fertilizers, among others, lead to several pollution issues that not only affect groundwater and soils, but also microbiota and surrounding wildlife, and of course, human health.

Therefore, it is essential to consider the implementation of products based on natural substances, whose characteristics may allow the substitution of chemical compounds in the agricultural sector. Biopolymers are one such possible solution to the problem because they are typically biodegradable materials obtained from renewable raw materials. Natural biopolymers include starch, cellulose, pectin, chitin, and chitosan, and have been part of humanity since its existence, being part of basic daily needs as fundamental as food and clothing, as well as medical materials, packaging, food additives, engineering plastics, chemicals for water treatment, among many others [1, 2]. The biopolymers known as chitin (poly-N-acetylglucosamine), and its derivative chitosan (obtained by deacetylation of chitin) have achieved a prominent place in the development of applications related to the treatment of water [3], soil moisture control [4], the intelligent release of fertilizers [5], growth stimulant [6], an inducer of defense mechanisms in plants [7], antimicrobial [8–10]. Their usefulness in this fight is such that their applications related to the control of population growth should also be considered due to the potential applications they have for the controlled release of contraceptives [11] and spermicides [12], including the expectations of creating a non-hormonal female contraceptive with no side effects [13]. One of the main biological applications of chitosan, which was recently discovered, is in the field of gene delivery, due to its ability to interact with anionic DNA [10, 14, 15].

Thus, the aim of this document is to present an exhaustive revision of the chito-oligosaccharides derived mainly from chitosan and their potential application in agriculture and other areas. In recent years, many studies have investigated the effects of chito-oligosacharides on human health, for example, as immunological modulators, [16, 17] anticancer [18, 19], antidiabetic [20, 21], and antimicrobial compounds [22, 23]. Some factors that potentiate the effect of chito-oligosacharides are their lower molecular weight and higher degree of deacetylation [24, 25]. The chito-oligosacharides can have antimicrobial properties, can induce plant resistance, and can stimulate plant growth as well, which makes them a promising alternative for the agricultural sector.

2. Shrimp species as a source of waste materials of industrial interest

Shrimp is one of the most important fishery resources worldwide, due to its nutritional value and high demand, especially in developed countries such as the United States of America, Japan, and the European Union [26].

Shrimp (*Penaeus* sp.) belongs to the animal kingdom, Phylum *Arthropoda,* class Crustacea, order Decapoda, and genus *Penaeus*. Like all arthropods, shrimp body is divided into three big main regions: cephalothorax, abdomen, and telson. It possesses various appendages such as antennules, antennae, mandibles, maxillae, maxillipeds, and pereiopods. This crustacean is essentially constituted by two parts: the crustacean muscular part, which corresponds to 50% of total mass, and the cephalothorax exoskeleton, including the tail, which is equivalent to the other 50% [27]. The cephalothorax is not fully exploited by man, so it is separated and broadly known as the shrimp waste. In Mexico, until recently, these residues were thrown back into the ocean or used as a source of proteins for fattening foods. However, in countries such as Japan, Thailand, and Korea, these shrimp wastes have acquired a very important commercial relevance [28].

Shrimp represented the 2.4% of the worldwide fishery production in 2018, with 3.75 million metric tons in live weight, and an establishedpotential of 4.0 million metric tons for the year 2021 [29]. China is the leading country in shrimp production, with about 1.5 million metric tons of the total fishing (**Figure 1**).

In Latin America, Ecuador is the leading country in shrimp production, and it was expected that the production in this country reached 700,000 metric tons in the year 2021, making Ecuador the third main worldwide shrimp producer, only after China and Vietnam (**Figure 2**). Although Mexico suffered severe losses in 2013, the Mexican industry was able to recover its production in 2015. Besides, it was expected a higher growth, where 180,000 metric tons were expected to be reached by 2021 [30]. In Mexico, in terms of catch volumes, the shrimp occupied second place in the national fishery production, with a live weight volume of 155,281 tons in 2018 (**Figure 3**).

In recent years there has occurred overexploitation in all the commercial fishing species, including shrimp, because of uncontrolled growth fishery activities, either artisanal or industrial. Due to this problem, preventive measures have been implemented, including a temporally closed season in 2003, in which it was prohibited any fishing of shrimp species; this closed season was extended until 2017. Besides, it was also implemented a specific fishing season under the Mexican Standard NOM-009-PESC-1993. The productions of aquaculture farms show an exponential growth tendency, while open sea mats and bay fishery have kept a constant behavior (**Figure 4**).

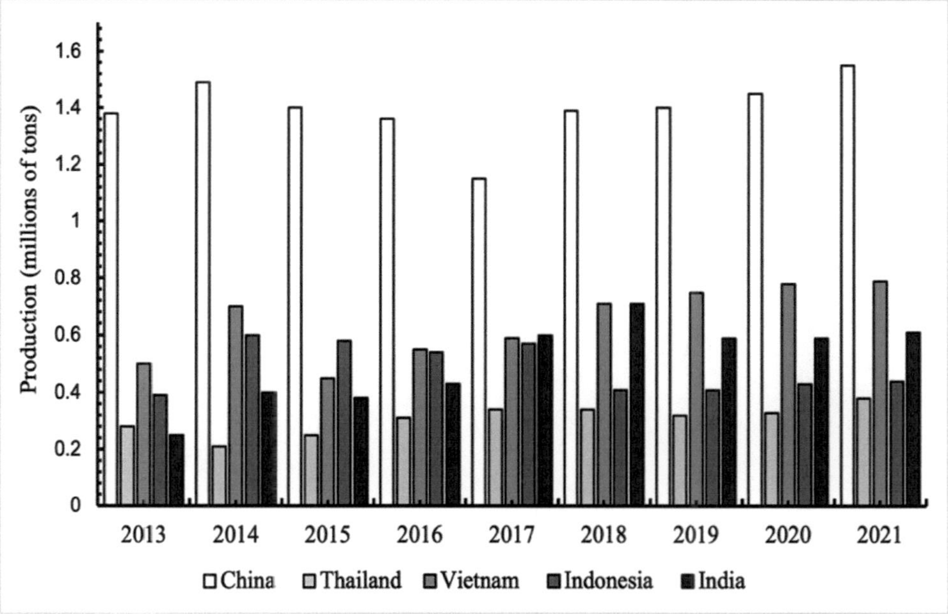

Figure 1.
Aquaculture shrimp production of the main producing countries. Sources: FAO [29] and Anderson et al. [30].

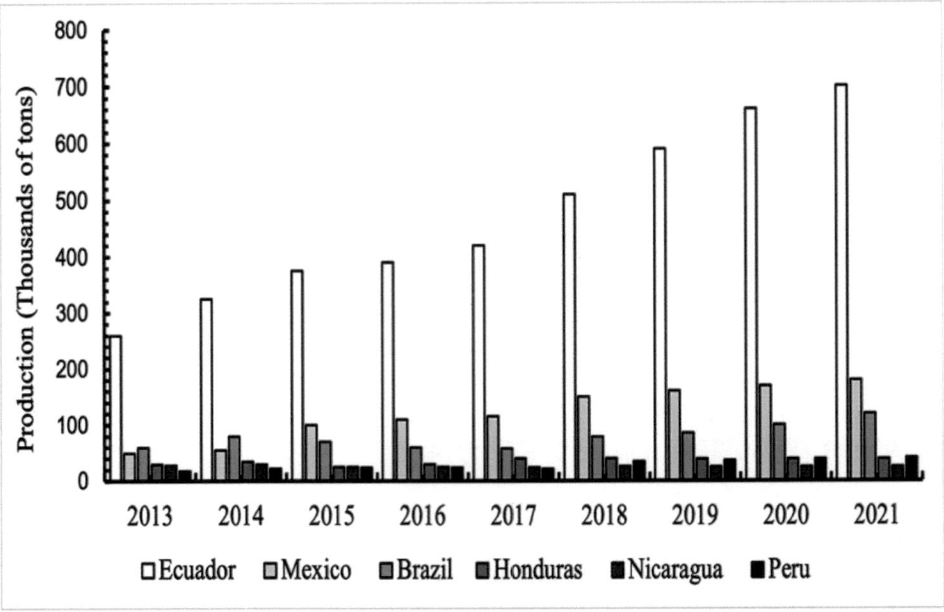

Figure 2.
Aquaculture shrimp production in the main producing countries of Latin America during 2013–2017 and projection for 2018–2021. Source: FAO [29] and Anderson et al. [30].

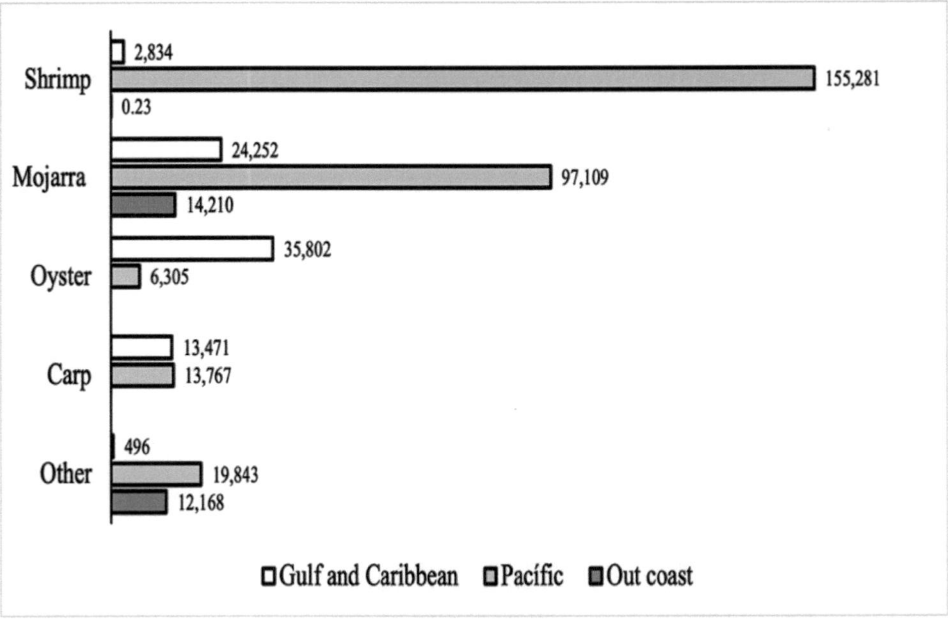

Figure 3.
Contribution of the main commercial species in the volume of national aquaculture production live weight in 2018. Adapted from the Mexican yearbook of aquaculture and fisheries, CONAPESCA [31].

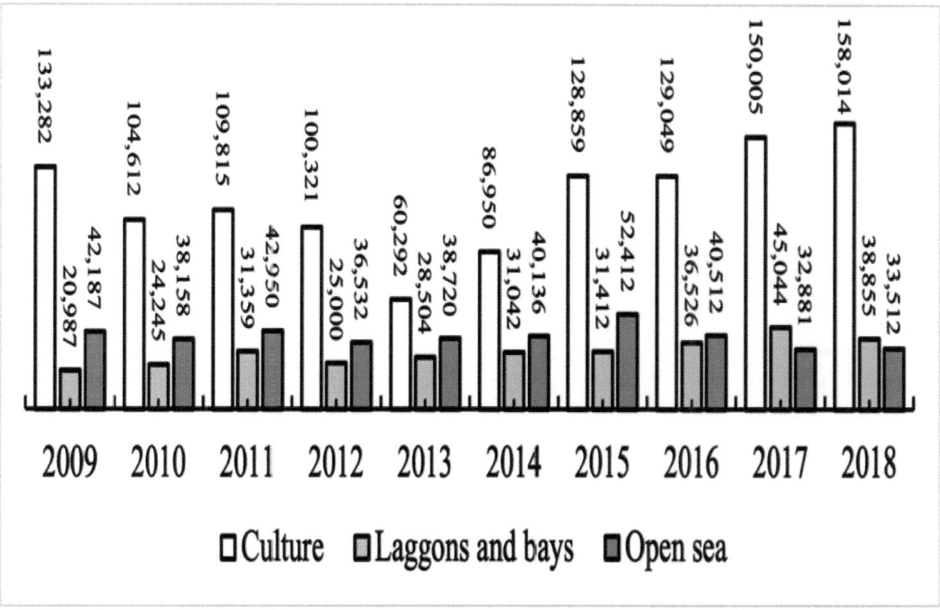

Figure 4.
Mexican live weight shrimp production by origin 2009–2018. Adapted from the Mexican yearbook of aquaculture and fisheries, CONAPESCA [31].

2.1 Current situation: the birth of Mexican industries based on the harnessing of shrimp wastes

Chitin and its derivates have commercial use in different countries, mainly Japan, China, and the USA. All these countries already have a consolidated industry around this polymer, being China the main producer and exporter of chitin. As to Chile and Spain, both countries are in an early stage, like the rest of Latin America. In Mexico, the interest in the chitin sector, as well as in chitosan, and their derivatives, is relatively new; therefore, both academic institutions and industries have manifested their interest in this topic. For example, scientists from Centro de Investigación en Alimentación y Desarrollo AC (CIAD) in Sonora, México [32], have informed that the head and exoskeleton of shrimp should be consider not as a waste but as source for chitin and chitosan production. Additionally, there are enterprises such as "Neptuno", where researchers and businessmen collaborate to produce these polysaccharides and their derivates. Until now, they have actively participated in conferences and presentations aiming to awaken the interest in this field. Another enterprise called "Polímeros acuícolas", from Guasave, Sinaloa, pretends to exploit shrimp wastes to produce substances of interest to satisfy the global need. Furthermore, some well-known Mexican enterprises, like Resistol and Comex, have already created links with the former company.

2.2 Chemical composition of shrimp wastes

Shrimp obtained from fishery is destined for direct human consumption; it can be found in different presentations that require diverse industrial processes, or also can

Author	[35]	[36]
% Water	7.87	4
% Dry matter	92.13	96
% Crude fiber	26.89	N/D
% Crude protein	34.5	46.3
& Fat	5.14	9.04
% Ashes	25.60	17.04
% Calcium	16.69	7.0
% Phosphates	0.85	3.03
Chitin energetic content	938*	2500
Chitin	18.7	9.82

**Calculated by the method of Schaibel (1980).*

Table 1.
Proximal analysis of the shrimp waste showing its content of chitin and energy (kcal/kg).

be consumed fresh or frozen. Considering that almost 50% of the crustacean weight is thrown before their consumption, it was calculated that the production of shrimp wastes in México will reach 79,138 tons in 2018 [31].

The crustacean exoskeletons are formed by successive protein-chitinous layers, with a high calcium carbonate content. Depending on the species, the chitin content can vary from 0.01 to 40% on a dry base, while protein content can fluctuate between 50 and 80%. Through X-ray diffraction studies, it has been determined that the polysaccharide chains of chitin in crustaceans, specifically in shrimp species, are tied in an antiparallel way, given a crystalline structure called type β [33]. Protein molecules adopt an antiparallel conformation of folded chains, while chitin molecules are arranged in a perpendicular way with respect to the protein chains, resulting in a tridimensional reticular structure, organized in layers with high mechanic strength. Therefore, such different protein-chitinous layers can adopt a "sandwich" structure, in whose center would be a nucleus made by chitin molecules, surrounded by layers of fibrous proteins arranged transversally [34]. In addition, it is important to emphasize that there are factors that can produce significant changes in the percentual composition of shrimp wastes, such as the crustacean species, season and geographical region of capture, and the storage of samplesas well. In general, there can be considered that such wastes contain an average of 34.5% protein, 26.89% crude fiber, and 25.60% ashes. More detailed data is shown in **Table 1**, where can also be seen that they possess a high calcium carbonate content and various phosphates. Therefore, these kinds of wastes mayrepresent one of the possible non-conventional protein sources for animals, with a good nutritional potential, because such proteins are especially rich in lysine, which could equilibrate diets based on cereals [37].

3. Chemical structure of chitin, chitosan, and their derivative chito-oligosaccharides

3.1 Chemical structure and properties

Chitin is a non-linear polymer constituted by units of N-acetyl-2-amino-2-deoxy-D-glucose (N-acetyl-glucosamine or NAG), joined by glycosidic bonds β $(1 \rightarrow 4)$ [38, 39].

DOI: http://dx.doi.org/10.5772/intechopen.106501

The conformation of such glycosidic bonds produces an alternative spatial location of the N-acetyl groups along the polymer chain, where the N-acetyl-glucosamine dimer chitobiose can be considered as the minimal structural repeating unit of chitin (**Figure 5**) [34]. On the other hand, chitin is one of the most abundant natural polymers, like cellulose and hemicellulose [40]. In nature, chitin is forming cover structures of arthropods, insects, arachnids, mussels, fungi, and some algae [41].

Concerning chitosan, this is also a lineal polysaccharide that could be obtained after the extensive deacetylation of chitin and, therefore, it is composed of two different aminated monosaccharides, which are randomly placed along the polymeric chain. Such monosaccharides are amino sugars NAG and D-glucosamine (GA), which are linked, likewise, by glycosidic bonds β (1 → 4), (**Figure 6**). It is important to point out that the total deacetylation of chitin is a quite complicated process, and therefore is possible to generally obtain mixtures of chitosans with different degrees of deacetylation (generally higher than 45%); thus, the criteria used to differentiate them is mainly their solubility in aqueous diluted acidic solutions [38].

3.2 Chitosan depolymerization to produce chito-oligosaccharides

Due to the very high molecular weight of chitosan, and its high viscosity as well, the use of this polymer becomes difficult for some applications. Therefore, this problem was solved by using the resulting products of its hydrolysis. However, such substances could be harder to obtain in the amounts required for large-scale industrial processes. In this regard, it has been reported that the hydrolysis or depolymerization of chitosan can be done through different methods: physical (ultrasound), chemical (hydrolytic reactions), or biological (using hydrolases). Among them, chemical hydrolysis is the most used at an industrial scale.

3.2.1 Chemical obtention of chito-oligosaccharides from chitosan

Chitosan can be hydrolyzed chemically either through acidic depolymerization or by an oxidative-reductive treatment. The acidic depolymerization is carried out by using a variety of chemicals such as hydrochloric, hydrofluoric, nitrous, sulfuric, and acetic acids. However, the use of such chemicals brings disadvantages, as their low yields obtained. Although they are relatively fast and cheap processes, they are inconvenient for their commercialization due to the production of toxic compounds and their considerable risk for the environment, since the materials used are highly residual [38]. There are multiple reports of chitosan oligomers obtained in this way and used alternatively in agriculture [43–46].

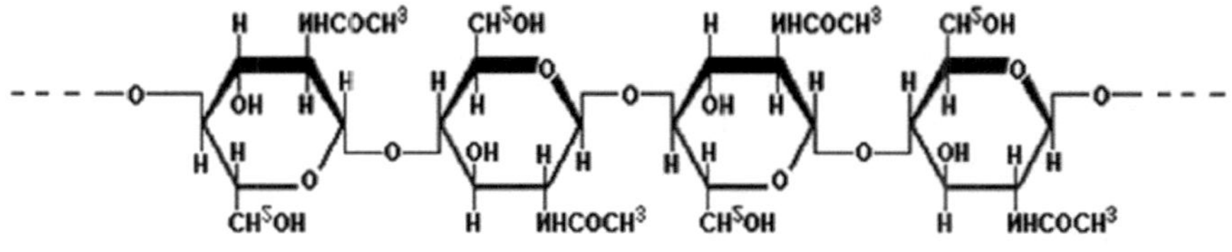

Figure 5.
Chemical structure of chitin.

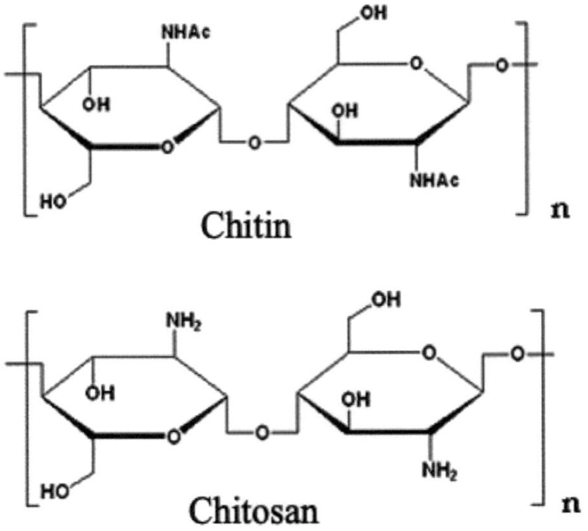

Figure 6.
Comparative chemical structures of chitin and chitosan [42].

3.2.2 Enzymatic obtention of chitosan oligosaccharides

Such enzymatic processes are carried out generally in discontinuous reactors and are preferred over the chemical methods, due to the reduction of adverse factors. These processes require specific enzymes like chitosanases or some less specific ones like cellulases, lipases, hemicellulases, and pectinases [47]. Chitosanases are enzymes broadly distributed in nature, which are capable of degrading chitosan into low molecular weight oligomers. These enzymes have been found in bacteria, viruses, fungi, and plants [48, 49]. However, there is a limitation in the use of these specific enzymes, because of their high cost and low availability in high amounts [50]. Due to this, some researchers have explored the use of non-specific commercial enzymes, which showed to be able to degrade chitosan, almost with the same efficiency than chitosanases, but cheaper [51]. As occurred with specific chitosanases, hydrolases are also able to catalyze the breakdown of β- (1,4)-glycosidic bonds present in chitosan. These enzymes have been found in microorganisms like viruses, bacteria, and fungi. During a recent research performed by Olicón-Hernández *et al.* [52], the extracellular chitosanase from *Bacillus thuringiensis*, grown in a chitosan containing medium, was used as a crude enzyme previously sterilized by filtration to produce a mixture of mono-, di-, tri-, and tetra-saccharides using colloidized chitosan as the substrate. These results are encouraging because they show that it is possible to transform chitosan obtained from shrimp wastes, through a microbial process, into products with biotechnological importance, avoiding the traditional use of chemical substances, as different acids.

4. Chitin and chitosan uses

Chitin can be used in a variety of fields. For instance, it was studied as a wound healing and blood thinner for medical purposes. It was also used as stationary support for the enzyme immobilization in column chromatography, due to its gelling and adhesive properties. In pharmacy is used as an excipient and dispenser of drugs. Besides, chitin has applications as an adhesive in textile and paper industries, and in agriculture for soil improvement. Due to its chelating properties, chitosan is used in water treatment and also for the decontamination of effluents. On the other hand, chitosan possesses physicochemical, functional, and biological properties, being useful in different fields such as medicine, pharmacy, agriculture, and food industry, among others. Chitosan has a high capacity for the sequestration of metallic ions, which is useful for the decontamination of industrial wastewater. Its polycationic nature grants a flocculant action, being also a good support for enzyme and cell immobilization, both in biotechnology and food industry [53].

Chitosan is also an excellent former of fibers, films, and membranes [54], and can be used to prepare microspheres or microcapsules; these capabilities, along with its biocompatibility and biodegradability, allow its use in both biomedical and pharmaceutic industries [55]. Also, it has been studied the use of chitosan as an excipient, to propitiate the controlled release of drugs and to reduce the cholesterol levels [56] and as a boost for the immune system and for the elaboration of the gels used in cosmetology [57]. Likewise, it has been described its antimicrobial action against pathogens and microorganisms that damage fruits and vegetables; this activity has been explained by supposing changes in the permeability of cells, due to the interactions between chitosan (a polycation) with the electronegative charges placed on the cell surface. Other therapeutical use concerns weight, high cholesterol, and burns controls [58]. Other applications, based on the polar capability of the hydroxyl and carboxyl groups of chitosan, have been proposed to make bio-electro sensors able to detect cancerogenic cells and at the same time being useful to administer antitumor agents to specific cells [59].

5. Some uses of chito-oligosaccharides

As indicated above, it is possible to obtain low-molecular-weight derivates of the above-cited polymers, by chemical or enzymatic treatments, which are known as chito-oligosaccharides, whose structure main contain scarcely 3 to 10 monosaccharide units. However, some authors consider that such chito-oligosaccharides may contain until 20 monosaccharide units in their chains. These compounds, particularly those derived from chitosan hydrolysis, can be used in a variety of biotechnological areas, standing out in medicine and pharmacy, due to their beneficial effects on human health, as is described hereafter [52].

5.1 Medical uses

Chito-oligosaccharides from chitosan confer an immunological modular effect, because theyit boost the immune system through cellular proliferation, besides possessing other stimulative immunological effects [57, 60–63]. In this particular instance, it is known that chito-oligosaccharides accelerate theformation of antibodies and also induce the cellular differentiation of leukocytes. Oligosaccharides from chitosan also possess an accelerating effect on intestinal transit since they help the proliferation of *Bifidobacterium* and *Lactobacillus* cells present in the intestinal flora [64].

Another important finding refers to their anti-tumoral activity, as it is believed that they can suppress and prevent cancer [65]. As metabolic emulators chito-oligosaccharides have anti-cholesterolemic effects, because they are able to reduce cholesterol, triglycerides and glucose levels in the blood; besides they reduce blood pressure and have anti-obesity effects.

As food additives, chito-oligosaccharides are used as dietary fiber, and dietary supplement for poultry species and livestock [63]. Other relevant applications for these compounds involve their use in arthritis control and as an antidiabetic; also in the treatment of gastric ulcers, as antimutages, anti-inflammatories and as low caloric sweeteners [66, 67].

On the other hand, it is important to point out that there are two factors that should be considered among the most relevant, related to the biological activity of chito-oligosaccharides; one is the length of their chains and the other is their deacetylation degree [68]. In general, the longer chito-oligosaccharides are, the stronger effects they may have. However, this does not imply that the shortest oligosaccharides do not possess similar (or other different) biological effects. For example, it has been observed that penta-, hexa- and hepta-glucosamines present the strongest and more varied biological effects [69].

5.2 Agricultural applications

It is known that chitin, and its derivates, have a broad range of biological activities which include antioxidant and, antimicrobial effects, and other properties that can be used on an industrial scale as well. Chitosan has been used for seed coverings, with the objective of controlling plagues and improving plant defense system against microorganisms [70]. It has been demonstrated that chito-oligosaccharides have diverse effects on plant cultivation and, on the enhancement of plant growth and development, besides improving both the quality and yield of the vegetable products [71]. In a study made by Mahdavi & Rahimi [70], it was tested the effect of chito-oligosaccharides in stimulation, specifically for germination and growth, of *Trachyspermum ammi*, observing that its growth was accompanied by a decrease in the damaging impact caused by abiotic stress like high salinity. Likewise, a study using three fractions of chito-oligosaccharides of different molecular weights, obtained from the same initial chitosan sample, clearly demonstrated that the fraction with the lowest molecular weight produced a higher acceleration in the germination of zucchini seeds covered with these compounds. In another study reported by Zou *et al.* [72], it was observed the significant benefits for soy yields, seed germination, and plant growth. Other uses of chito-oligosaccharides in the agricultural sector are as agents to conserve seeds [73], as plant defense enhancers and for the protection against plagues and diseases [74].

5.2.1 Effects on plant resistance to diseases

Jia *et al.* [75] used *Arabidopsis* plants which were pre-treated with 50 mg/L of chito-oligosaccharides per day, before their inoculation with the tobacco mosaic virus (TMV); it was found that the expression of defenses, associated with genes related to pathogenicity, resulted strengthened.

In another study, the efficiency of chito-oligosaccharides to prevent and control the southern rice striped black dwarf virus was demonstrated, evidencing that these compounds regulate the increase of proteins related to plant defenses. Also, a field test has been carried out after treatment with chito-oligosaccharides as an antifungal agent in grape plantations, where results showed that the mortality and infection rates were reduced significatively on inoculated plants with pathogens such as *Diplodiaseriata* y *Phaeomoniella chlamydospora* [76].

5.2.2 Effects on growth and plant development

The use of oligosaccharides can help to improve plant growth, seed germination, chlorophyll content, nitrogen fixation, and nutrient absorption. Oligomers from chitosan not only have the property of inducing resistance, but they can promote plant growth and development. During their interaction with tobacco cells, these compounds regulate concentrations of indole-3-acetic acid (IAA) and their related peroxidases, which indicates a growth accelerator effect. Also, field tests have been done to evaluate oregano growth using different oligosaccharide doses (50–1000 ppm),

Species	Properties of QOS*	Concentration used	Mode of use	Effects/ observations	Reference
Tomato	—	25, 50, 75, 100 mg/L	Foliar	Plants height, leaves number, fruit yield	[77]
Wheat	93% GD	15 mg/L	Foliar	Chlorophyll, activity PEPC, saccharose content, TCA cycle	[78]
Mint	90% GD	40, 60, 120, 160 mg/L	Foliar spray	Longer root length	[79]
Lemmon grass	90% GD	40, 60, 80, 100 mg/L	Foliar spray	Root length	[80]
Chili	91.4% GD; PM 8 kDa	50 mg/L	Foliar	Increase on chlorophyll content	[81]
Wheat	95% GD	0.01%	Foliar	Sprout length, root length, chlorophyll content, sugar content	[72]
Bean	85% GD	100 mg/L	Foliar	Plant height, more vanilla pods, seeds yield	[82]
Barley	Different PM	—	Plant tissue culture	Higher growth and seed yield	[83]
Lavender	30% GD; PM 16 kDa	30, 40, 100 mg/L	Plant tissue culture	Higher plant length, multiplication, and sprout weight.	[84]
Curly kale	—	10, 50, 100 mg/L	Hydroponic cultivation	Enhance of growth and reduction for harvesting time	[84]
Soy, wheat and rice	—	10–100 mg/L	Hydroponic cultivation	They promote growth under stress	[85]

**GD = degree of deacetylation, PM = molecular weight.*

Table 2.
Some reports of the effects of oligosaccharides on the physiological attributes of plants.

where results indicated that, from 200 to 500 ppm chito-oligosaccharides, there was an increase in plant height, while doses from 50 to 200 ppm significantly regulated the concentration of polyphenols.

5.2.3 Effects on the quality of vegetable products

Studies carried out in this regard showed that treatment with chito-oligosaccharides can significantly improve the quality of strawberry plants; in such research a treatment with 50 mg/L of these chito-oligosaccharides, applied to the fruits prior to harvest them, increased their pulp viscosity, lignin content, sugar, protein, and titratable acidity; besides, they strengthen the strawberry antioxidant capability due to a higher production of components such as anthocyanins, total phenols, flavonoids, and vitamin C. **Table 2** shows some works concerning the effects of chito-oligosaccharides on diverse physiological attributes of plants.

6. Conclusions

Sustainable agriculture is a relevant topic considering the growing population worldwide and the severe damage that chemicals have caused to the environment due to their use for agricultural purposes. Many agrochemicals used nowadays are costly in the global market; however, they promote agricultural production. There is well known that many of these compounds used to protect crops against diseases and to increase yields, are considered pollutants of soil, crops, biological diversity, microbiota and, in addition, may cause diseases in animals and humans. Thus, there is currently an enormous need to promote healthier ecosystems and support sustainable soil management to minimize the use of these harmful synthetic agrochemicals, while promoting the development of methods and products for the control of pests and diseases, that could be more respectful to the environment. The utilization of protein-chitinous residues, which are abundant in our country (and without the correct management, may also generate pollution to the environment), for the obtention of derivates like chitin, chitosan, and chito-oligosaccharides from chitosan, among others, are considered valuable in many fields, due to their properties like biocompatibility, biodegradability, and null toxicity. Therefore, they are an alternative to conventional agriculture, being a feasible alternative to enhance productivity and crop protection. To that effect, chito-oligosaccharides may fulfill environmental and health requirements to help meet the needs of a constantly growing population, which involve the production of high-quality food along with low environmental impact. Unlike chitin and chitosan, chito-oligosaccharides are soluble in water, and due to their low molecular weight, they can enter the cell and have a higher biological effectivity, not to mention their relevant capability to stimulate seed germination, plant growth, to activate resistance mechanisms in crops, antimicrobial activity, and more important, they can be obtained by microbial methods that are harmless to the environment and at a low cost.

Acknowledgements

We are deeply grateful for the editor of Chitin-Chitosan Isolation, Properties, and Applications. We are also grateful to IntechOpen and Universidad Veracruzana which supported our research in various ways.

Conflict of interest

The authors declare no conflict of interest.

Author details

Paul Edgardo Regalado-Infante[1], Norma Gabriela Rojas-Avelizapa[2], Rosalía Núñez-Pastrana[1], Daniel Tapia-Maruri[3], Andrea Margarita Rivas-Castillo[4], Régulo Carlos Llarena-Hernández[1] and Luz Irene Rojas-Avelizapa[1*]

1 Faculty of Biological and Agricultural Sciences, Universidad Veracruzana, Veracruz, Mexico

2 Research Center in Applied Science and Advanced Technology, CICATA-Instituto Poitécnico Nacional, Qro, Querétaro, México

3 Biotic Product Development Center, CEPROBI-Instituto Politécnico Nacional, Yautepec, Morelos, México

4 Technological University of the Metropolitan Zone of the Valley of Mexico UTVAM Tizayuca, Hidalgo, México

*Address all correspondence to: luzrojas@uv.mx

References

[1] Hill JW, Kolb DK. Chemistry for Changing Time. 14th (ed) ed. Vol. 816. Malaysia: Prentice Hall; 2015

[2] Marsh K, Bugusu B. Food packaging-roles, materials, and environmental issues. Journal of Food Science. 2007;**72**:39-55

[3] Lárez-Velásquez C, Millán-Barrios E. Chitosan for pesticide control on environmental protection and water purification. In: Gertsen N, Sonderby L, editors. Series: Air, Water and Soil Pollution Science and Technology. Chapter 4 ed. New York, United States of America: Nova Publisher; 2009. ISBN: 978-1-60741-599-2

[4] Sabadini R, Martins V, Pawlicka A. Synthesis and characterization of gellan gum: Chitosan biohydrogels for soil humidity control and fertilizer release. Cellulose. 2015;**22**(3):2045

[5] Pereira A, Martins A, Paulino A, Fajardo A, Guilherme M, Iecher-Faria M, et al. Recent advances in designing hydrogels from chitin and chitin–derivatives and their impact on environment and agriculture: A review. Revista Virtual de Química. 2017;**9**:17

[6] Malerba M, Cerana R. Recent advances of chitosan applications in plants. Polymers. 2018;**10**(2):118. DOI: 10.3390/polym10020118

[7] Xing K, Zhu X, Peng X, Qin S. Chitosan antimicrobial and eliciting properties for pest control in agriculture: A review. Agronomy for Sustainable Development. 2015;**35**(2):569-588. DOI: 10.1007/s13593-014-0252-3

[8] Kong M, Chen X, Xing K, Park H. Antimicrobial properties of chitosan and mode of action: A state of theart review. International Journal of Food Microbiology. 2010;**144**(1):51-63. DOI: 10.1016/j.ijfoodmicro.2010.09.012

[9] Guzmán K, Kumar B, Vallejo M, Grijalva M, Devut A, Cumbal L. Ultrasound- assisted syntesis and antibacterial activity of gallic acid-chitosan modified silver nanoparticles. Progress in Organic Coatings. 2019:1-8. DOI: 10.1016/j.porgcoat.2019.01.009

[10] Kumar S, Deepak V, Kumari M, Dutta PK. Antibacterial activity of diisocyanate- modified chitosan for biomedical applications. International Journal of Biological Macromolecules. 2016;**84**:349-353. DOI: doi.org/10.1016/j.ijbiomac.2015.12.027

[11] Agrawal S, Pruthi J. Development and evaluation of matrix type transdermal patch of ethinyl–estradiol and medroxyprogesterone acetate for anti–implantation activity in female Wistar rats. Contraception. 2011;**84**(5):533. DOI: 10.1016/j.contraception.2011.03.005

[12] Smith R. Chitosan as a Contraceptive. Patent No.: US 4,474,769 A. 1984

[13] Kootala S, Filho L, Srivastava V, Linderberg V, Moussa A, Trombotto H, et al. Reinforcing mucus barrier properties with low molar mass chitosans. Biomacromolecules. 2018;**19**(3):872. DOI: 10.1021/acs.biomac.7b01670

[14] Kumar S, Koh J. Physiochemical and optical study of chitosan-terephthaldehyde derivative for biomedical applications. International Journal of Biological Macromolecules. 2012;**51**:1167-1172. DOI: 10.1016/j.ijbiomac.2012.09.001

[15] Kumar S, Garg P, Pandey S, Kumari M, Hoon S, Jang K, et al. Enhanced chitosan–DNA interaction by 2-acrylamido-2-methylpropane coupling for an efficient transfection in cancer cells. Journal of Materials Chemistry B. 2015;**3**:3465-3475. DOI: doi.org/10.1039/C4TB02070G

[16] Bahar B, O'Doherty JV, Maher S, McMorrow J, Sweeney T. Chitooligosaccharide elicits acute inflammatory cytokine response through AP-1 pathway in human intestinal Epitheliallike (Caco-2). Cellular & Molecular Immunology. 2009;**51**(3):283-291. DOI: 10.1016/j.molimm.2012.03.027

[17] Walsh AM, Sweeney T, Bahar B, Flynn B, O'Doherty JV. The effect of Chitooligosaccharide supplementation on intestinal morphology, selected microbial populations, volatile fatty acid concentrations and immune gene expression in the weaned pig. Animal. 2012;**6**(10):1620-1626. DOI: 10.1017/S1751731112000481

[18] Shen KT, Chen MH, Chan HY, Jeng JH, Wang YJ. Inhibitory effects of Chitooligosaccharides on tumor growth and metastasis. Food and Chemical Toxicology. 2009;**47**(8):1864-1871. DOI: 10.1016/j.fct.2009.04.044

[19] Kim EK, Je JY, Lee SJ, Kim YS, Hwang JW, Sung SH, et al. Chitooligosaccharides induce apoptosis in human myeloid leukemia HL-60 cells. Bioorganic & Medicinal Chemistry Letters. 2012;**22**(19):6136-6138

[20] Ju C, Yue W, Yang Z, Zhang Q, Yang X, Liu Z, et al. Antidiabetic effect and mechanism of Chitooligosaccharides. Biological and Pharmaceutical Bulletin. 2010;**33**(9):1511-1516. DOI: 10.1248/bpb.33.1511

[21] Karadeniz F, Artan M, Hong C, Kim S. Chitooligosaccharides protect pancreatic b-cells from hydrogen peroxide-induced deterioration. Carbohydrate Polymers. 2010;**82**(1):143-147

[22] Quintero-Villegas M, Aam B, Rupnow J, Sorlie M, Eijsink V, Hutkins R. Adherence inhibition of Enteropathogenic *Escherichia coli* by Chitooligosaccharides with specific degrees of acetylation and polymerization. Journal of Agricultural and Food Chemistry. 2013;**61**(11):2748-2754. DOI: 10.1021/jf400103g

[23] Xia W, Liu P, Zhang J, Chen J. Biological activities of chitosan and Chitooligosaccharides. Food Hydrocolloids. 2011;**25**(2):170-179

[24] Benhabiles M, Salah R, Lounici H, Drouiche N, Goosen M, Mameri N. Antibacterial activity of chitin, chitosan and its oligomers prepared from shrimp Shell waste. Food Hydrocolloids. 2012;**29**(1):48-56

[25] Fernandes J, Eaton P, Nascimento H, Giao M, Ramos O, Belo L, et al. Antioxidant activity of chitooligosaccharides upon two biological systems: Erythrocytes and bacteriophages. Carbohydrate Polymers. 2010;**79**:1101-1106. DOI: org/10.1016/j.carbpol.2009.10.050

[26] Caso M, Pisanty I, Ezcurra E. Diagnostico ambiental del golfo de México. In: Ecosur, editor. Secretaría del Medio Ambiente y Recursos Naturales. Vol. 2. México: Instituto Nacional de Ecología; 2009. pp. 899-936

[27] Carranco M, Calvo C, Arellano L, Pérez-Gil L, Ávila E. Inclusión de la harina de cabezas de camarón *Penaeus* sp. en raciones para gallinas ponedoras. Efecto sobre la concentración de pigmento rojo de yema y calidad de huevo. Interciencia. 2003;**28**:328-333. ISSN 0378-1844

[28] Rojas-Avelizapa L, Cruz-Camarillo R, Guerrero M, Rodríguez-Vázquez R, Ibarra J. Selection and characterization of a proteo-chitinolytic strain of *bacillus thuringiensis*, able to grow in shrimp waste media. World Journal of Microbiology and Biotechnology. 1999;**15**:261

[29] FAO. El Estado Mundial de la Pesca y la Acuacultura 2020. La sostenibilidad en acción Roma; 2020. DOI: 10.4060/ca9229es

[30] Anderson J, Valderrama D, Darryl EGOAL. Revisión de la producción mundial de camarones. Global Aquaculture Alliance. 2019;**2019**:1-5

[31] Anuario de acuacultura y pesca 2018 [internet]. 2018. Available from: https://www.conapesca.gob.mx/work/sites/cona/dgppe/2018/ANUARIO_2018.pdf

[32] Centro de Investigación en Alimentación y Desarrollo A.C. 2008 [internet]. 2008. Available from: http://www.ciad.mx/content/view/173/1.

[33] Kurita K. Chitin and chitosan: Functional biopolymers from marine crustaceans. Marine Biotechnology. 2006;**8**:203-226. DOI: 10.1007/s10126-005-0097-5

[34] Cruz-Camarillo R, Rojas-Avelizapa L. Las quitinasas bacterianas y sus posibles aplicaciones biotecnológicas. In: Avances en purificación y aplicaciones de enzimas en biotecnología. México: UAM-Iztapalapa; 1999. pp. 238-324

[35] Mahata M, Dharma A, Rianto I, Rizal Y. Effect of substituting waste hydrolysate of *Penaeus merguensis* for fish meal in broiler performance. Pakistan Journal of Nutrition. 2008;**7**:806-810 ISSN 1680-5194

[36] Faniamo A, Oduguwa B, Oduguwa O, Ajasa O, Jegede O. Feeding value of shrimp meal for growing pigs. Archivos de zootecnia. 2004;**53**:77-85

[37] Okoye F, Ojewola G, Njoku-Onu K. Evaluation of shrimp waste meal as a probable animal protein source for broiler chicken. International Journal of Poultry Science. 2005;**4**:458-461. DOI: 10.3923/ijps.2005.458.461

[38] Castañeda R, Fuente S, Pacheco R, Ortiz-Rodríguez T, Barboza J. Potencial de los quito-oligosacáridos generados de quitina y quitosana. Acta Universitaria. 2014;**21**:14-23. DOI: 10.15174/au.2011.16

[39] Zhu X, Zhou Y, Feng J. Analysis of both chitinase and chitosanase produced by *Sphingomonas* sp. CJ-5. Journal of Zhejiang University Science B. 2017;**8**:831-838. DOI: 10.1631/jzus.2007.B0831

[40] Roberts G. Truth or myth Euchis. Newsletter. 2006;**21**:5-6

[41] Mukhtar K, Khan M, Siddiqui H, Jahan A. Chitosan and its oligosaccharides, a promising option for sustainable crop production- a review. Carbohydrate Polymers. 2019;**115**:1-64. DOI: 10.1016/j.carbpol.2019.115331

[42] Ghanbarzadeh B, Almasi H. Biodegradable Polymers. Chapter 6: Life of Science. IntechOpen Science; 2013. pp. 141-185. DOI: 10.5772/56230

[43] Dzung N. Enhancing crop production with chitosan and its derivatives. In: Kim S-K editor. Chitin, Chitosan, Oligosaccharides and their Derivatives. Vol. 14. Department of Physics, Korea Advanced Institute of Science and Technology: CRC Press; 2010. pp. 619-631. DOI: 10.1201/EBK1439816035-c42

[44] Katiyar D, Hemantaranjan A, Singh B, Bhanu N. Future perspective in crop protection: Chitosan and its oligosaccharides. Advances in Plants &

Agriculture Research. 2014;**1**:2-8. DOI: 10.15406/apar.2014.01.00006

[45] Ahmed K, Khan M, Siddiqui H, Jahan A. Chitosan and its oligosaccharides, a promising option for sustainable crop production - a review. Carbohydrate Polymers. 2020;**227**:115-331. DOI: 10.1016/j.carbpol.2019.115331

[46] Jia X, Rajib M, Yin H. Recognition pattern, functional mechanism and application of chitin and chitosan oligosaccharides in sustainable agriculture. Current Pharmaceutical Design. 2020;**26**:3508-3521. DOI: 10.2174/1381612826666200617165915

[47] Abdel-Aziz S, Kahil T, Keera A. Kinetic behavior of free and in situ immobilized chitosanases produced by the fungus *Mucor rouxii*. World Applied Sciences Journal. 2014;**30**:01-09. DOI: 10.5829/idosi.wasj.2014.30.01.13980

[48] Zhang P, Zhou W, Wang P, Wang L, Tang M. Enhancement of chitosanase production by cell immobilization of *Gongronella* sp. JG. Brazilian Journal of Microbiology. 2013;**44**:189-195. DOI: 10.1590/S1517-83822013005000017

[49] Sinha S, Tripathi P, Chand S. New bifunctional chitosanase enzyme from *Streptomyces* sp. and its application in production of antioxidant chito-oligosaccharides. Applied Biochemistry and Biotechnology. 2012;**167**:1029-1039. DOI: 10.1007/s12010-012-9546-6

[50] Rodriguez-Herrera R, Contreras-EsquivelJ,Mauricio-BenavidesJ, Garza-Garcia Y, Charles-Rodriguez A. Chitosanase production by a new bacterial source. Research Journal of Biological Sciences. 2008;**3**:957-963

[51] Struszczyk K, Szczęsna-Antczak M, Walczak M, Pomianowska E, Wojciechowska J. Enzymatic preparations from *Mucor moulds* and their application in oligoaminosaccharides production. Progress on Chemistry and Application of Chitin and its Derivatives. 2009;**14**:89-100. DOI: 10.15259/PCACD.21.09

[52] Olicón-Hernández D, Vázquez-Landaverde P, Cruz-Camarillo R, Rojas-Avelizapa L. Comparison of chito-oligosaccharides production from three different colloidal chitosans using the endochitonsanolytic system of *bacillus thuringiensis*. Preparative Biochemistry and Biotechnology. 2017;**47**:116-122. DOI: 10.1080/10826068.2016.1181086

[53] Lathouder K, van-Benthem T, Wallin S, Moulijn J. Polyethyleneimine (PEI) functionalized ceramic monoliths as enzyme carriers: Preparation and performance. Journal of Molecular Catalysis B Enzymatic. 2008;**50**:20-27. DOI: 10.1016/j.molcatb.2007.09.016

[54] Wittaya-areekul S, Prahsarn C, Sungthongjeen S. Development and *in vitro* evaluation of chitosan- Eudgragit RS 30D composite wound dressings. AAPS PharmSciTech. 2006;**13**(3): 123-128. DOI: 10.1016/j.ijpharm.2006.01.027

[55] Sezer A, Cevher E, Hatipoglu F, Ogurtan Z, Bas A. Preparation of fucoidan-chitosan hydrogel and its application as burn healing accelerator on rabbits. Biological and Pharmaceutical Bulletin. 2008;**31**:2326-2333. DOI: 10.1248/bpb.31.2326

[56] Zhao W, Yin Y, Liang Q. Controlled released of salidroside from chitosan-alginate microcapsules. Food Fermentation Industries. 2004;**30**:66-69. DOI: 10.1002/jps.2600830213

[57] Li X, Piao X, Kim S, Liu P, Wang L. Effects of chito-oligosaccharides

supplementation on performance nutrient Digestibilly, and serum composition in broiler chickens. Poultry Science. 2007;**86**:1107-1114. DOI: 10.1093/ps/86.6.1107

[58] Sugano M. A novel use of chitosan as a hypocholesterolemic agent in rats. The American Journal of Clinical Nutrition. 1980;**33**:787-793

[59] Wu J, Tsai G. Cellulase degradation of shrimp chitosan for the preparation of a water-soluble hydrolysate with immunoactivity. Fisheries Science. 2004;**70**:1113-1120. DOI: 10.1111/j. 1444-2906.2004.00912.x

[60] Suzuki K, Mikami T, Okawa Y, Tokoro A, Suzuki S. Antitumor effect of hexa-N-acetylchitohexaose and chitohexaose. Carbohydrate Research. 1986;**151**:403-408

[61] Shibata Y, Foster L, Metzger W, His-Myrvik Q. Alveolar macrophage priming by intravenous administration of chitin particles, polymers of N-acetyl-D-glucosamine, in mice. Infection and Immunity. 1997;**65**:1734-1741. DOI: 10.1128/iai.65.5.1734-1741.1997

[62] Hsiao Y, Lin Y, Su C, Chiang B. High degree polymerized chito-oligosaccharides synthesis by chitosanase in the bulk aqueous system and reversed micellar microreactors. Process Biochemistry. 2008;**43**:76-82. DOI: 10.1016/j.procbio.2007.10.017

[63] Dalian Haixin Chemical Industry Co. [internet]. Available from: http://www.ec21.com/offer_details/sell_chitosan_oligosaccharide--8054608.html

[64] Hyean-Hoo L, Yoon-Sun P, Jong- SJ, Woon-Seob S. Chitosan oligosaccharides, dp 2-8, have prebiotic effect on the *Bifidobacterium bifidum* and *lactobacillus* sp. Anaerobe. 2002;**1**:319-324. DOI: 10.1016/S1075-9964(03)00030-1

[65] Fernandes J, Sereno J, Garrido P, Parada B, Cunha M. Inhibition of bladder tumor growth by Chito-oligosaccharides in an experimental carcinogenesis model. Marine Drugs. 2012;**10**:661-1675. DOI: 10.3390/md10122661

[66] Liu B, Wan-Shun L, Bao-Qin H, Yu-Yin S. Antidiabetic effects of chito-oligosaccharides on pancreatic islet cells in streptozotocin-induced diabetic rats. World Journal of Gastroenterology. 2007;**13**:725-731. DOI: 10.3748/wjg.v13.i5.725

[67] Nam K, Kim M, Shon YH. Chemopreventive effect of chitosan oligosaccharide against colon carcinogenesis. Journal of Microbiology and Biotechnology. 2007;**17**:1546-1549

[68] Choi Y, Kim E, Piao Z, Yun Y, Shin Y. Purification and characterization of chitosanase from bacillus sp. strain KCTC 0377BP and its application for the production of chitosan oligosaccharides. Applied Environmental Microbiology. 2004;**70**:4522-4531. DOI: 10.1128/AEM.70.8.4522-4531.2004

[69] Zhang N, Andresen B, Zhang C. Inflammation, and reactive oxygen species in cardiovascular disease. World Journal of Cardiology. 2010;**2**:408-410. DOI: 10.4330/wjc.v2.i12.408

[70] Mahdavi B, Rahimi A. Seed priming with chitosan improves the germination and growth performance of ajowan (*Carum copticum*) under salt stress. Eurasian Journal of Biosciences. 2013;**7**:69-76

[71] Larez-Velasquez C, Chirinos A, Tacoronte M, Mora A. Chitosan oligomers as bio–stimulants to zucchini (*Cucurbita pepo*) seeds germination. Agricultures (Poľnohospodárstvo). 2012;**58**:113-119

[72] Zou P, Tian X, Dong B, Zhang C. Size effects of chitooligomers with certain degrees of polymerization on the chilling tolerance of wheat seedlings. Carbohydrate polymers. 2017;**160**:194-202. DOI: 10.1016/j.carbpol.2016.12.058

[73] Abu-Hassan M, Pei-Li T, Noor Z. Coagulation and flocculation treatment of wastewater in textile industry using chitosan. Journal of Chemical and Natural Resources Engineering. 2009;**4**:43-53

[74] Dutta P, Dutta J, Tripathi V. Chitin and chitosan: Chemistry, properties, and applications. Journal of Scientific & Industrial Research. 2004;**63**:20-31

[75] Jia X, Meng Q, Zeng H, Wang W, Yin H. Chitosan oligosaccharide induces resistance to *tobacco* mosaic virus in Arabidopsis via the salicylic acid-mediated signalling pathway. Nature. 2016;**6**:26144. DOI: 10.1038/srep26144

[76] Cobos R, Mateos R, Alvarez JM, Olego M, Sevillano S. Effectiveness of natural antifungal compounds in controlling infection by grapevine trunk disease pathogens through pruning wounds. Applied Environmental Microbiology. 2015;**81**:18. DOI: 10.1128/AEM.01818-15

[77] Monirul I, Humayun K, Mamun A, Pronabananda D. Studies on yield and yield attributes in tomato and chilli using foliar application of oligo-chitosan. GSC Biological and Pharmaceutical Sciences. 2018;**3**:20-28. DOI: 10.30574/gscbps.2018.3.3.0038

[78] Zhang X, Li K, Xing R, Liu S, Chen X. miRNA and mRNA expression profiles reveal insight into chitosan-mediated regulation of plant growth. Journal of Agricultural and Food Chemistry. 2018;**66**:3810-3822. DOI: 10.1021/acs.jafc.7b06081

[79] Ahmad B, Khan M, Jaleel H, Sadiq Y, Shabbir A. Exogenously sourced γ-irradiated chitosan-mediated regulation of growth, physiology, quality attributes and yield in *Mentha piperita* L. Turkish Journal of Biology. 2017;**41**:388-401. DOI: 10.3906/biy-1608-64

[80] Jaleel H, Khan M, Ahmad B, Shabbir A, Sadiq Y. Essential oil and Citral production in field-grown lemongrass in response to gamma-irradiated chitosan. Journal of Herbs, Spices & Medicinal Plants. 2017;**23**:378-392. DOI: 10.1080/10496475.2017.1349702

[81] Dzung P, Phu D, Du B, Ngoc L, Duy N. Effect of foliar application of oligochitosan with different molecular weight on growth promotion and fruit yield enhancement of chili plant. Plant Production Science. 2017;**20**:389-395. DOI: 10.1080/1343943X.2017.1399803

[82] El-Sawy N, El-Rehim H, Elbarbary A, Hegazy S. Radiation-induced degradation of chitosan for possible use as a growth promoter in agricultural purposes. Carbohydrate Polymers. 2010;**79**:555-562. DOI: 10.1016/j.carbpol.2009.09.002

[83] Luan L, Nagasawa N, Tamada M, Nakanishi T. Enhancement of plant growth activity of irradiated chitosan by molecular weight fractionation. Radioisotopes. 2006;**55**:21. DOI: 10.3769/radioisotopes.55.21

[84] Hafiz M, Muhamad O, Muhammad S, Ahmad M. Irradiated chitosan as plant promoter. Proceedings. Symposium Biological Gunaan. 2003; **3-4**:1-7. DOI: 10.1063/1.4940072

[85] Tham L, Nagasawa N, Matsuhashi S, Ishioka N, Ito T. Effect of radiation-degraded chitosan on plants stressed with vanadium. Radiation Physics and Chemistry. 2001;**61**:171-175. DOI: 10.1016/j.jrras.2015.03.008

Chapter 2

Preparation of Water-Soluble Colloidal Chitin (WSCC) from Prawn Waste and Its Characterization

Renuka Vinothkumar and Janet Paterson

Abstract

Chitin, the shell material of prawn, is a biodegradable polymer and environmentally biocompatible with low toxicity. Chitosan is the deacetylated form of chitin, which consists of poly-D-glucosamine units with no or few N-acetyl-D-glucosamine units. Commercial applications of these natural polymers are increasing in various sectors. Therefore, in addition to the environmental benefit, it may be economical to recover chitin from prawn waste. Chitosan is soluble in various organic acids, solvents and water. The poor solubility of chitin is the major limiting factor in its use in industrial applications. Number of studies have investigated to overcome the solubility problem of chitin. This research focuses on a new way of developing water-soluble colloidal chitin (WSCC) from prawn waste and investigates its fundamental rheological and antibacterial properties. WSCC films studied during this research may be used in food packaging or in medical applications. The use of WSCC biodegradable films will protect the environment in the future and will be an effective alternative to plastics that threatens the environment. The antibacterial study may be applied in pharmaceutical, medical and food packaging and coating applications. This research was conducted at the University of New South Wales, Australia in 2008.

Keywords: prawn waste, water-soluble chitin, chitin characterization, chitin rheology, antibacterial

1. Introduction

Pollution of soil, air and water contributed to environmental deterioration; its control is necessary. Plastics have become part of our lives; the treatment of waste plastics has become a serious problem because of the difficulty of land reclamation and disposal by incineration [1]. Recent interests have focused mainly on biodegradable plastics that are biocompatible to the environment [2]. Chitin and chitosan are examples of biodegradable, biorenewable and biofunctional polymers derived from seafood processing waste [3–8]. Therefore, in addition to the environmental benefit, it may be economical to recover chitin from prawn waste. The poor solubility of chitin is the major limiting factor in its use in industrial applications [9]. But chitosan is soluble in various organic acids, solvents and water [10]. Several studies have investigated the solubility problem of chitin. This research focuses on a new way of developing water-soluble colloidal chitin (WSCC) from prawn waste and investigates its fundamental rheological and antibacterial properties.

2. Literature review

2.1 Chitin and chitosan

Chitin is derived from the Greek word *chiton*, which means a coat of nail. Chitin is the major component of the exoskeleton of invertebrates and the cell wall of fungi and yeast [11, 12], from mushrooms [13]. Chitosan is usually obtained by alkaline or enzymatic deacetylation of chitin. The importance of chitin and chitosan has grown partly because they represent a renewable and biodegradable source of materials, and partly because of the recent increased understanding of their functionality in various applications [2, 14–25]. Usually, chitin is prepared from crustacean waste through deproteinization using alkali or enzyme, demineralization or decalcification using acid followed by decolourization using decolouring agents in order to remove the proteins, calcium and colour, respectively. The chitin thus obtained can then be deacetylated either by alkali or enzyme to produce chitosan [26]. The properties of chitin and chitosan depend on the processing conditions. Chitosan prepared from chemical and enzymatic deacetylation of chitin differs in their degree of deacetylation (%), distribution of acetyl groups, chain length and conformational structure of chitin and chitosan molecules. These factors affect the characteristics of chitin and chitosan [27].

2.2 Solubility of chitin and its derivatives in water

The most remarkable difference between chitin and chitosan is their solubility. Chitin is insoluble in almost all solvents; chitosan dissolves in almost all aqueous acids. The insolubility of chitin is the major disadvantage to its use. Solvents of chitosan are generally safe to consume, thus allowing its use in various industries including the food industry. Most solvents used for the dissolution of chitin are toxic, hence, they cannot be used in food processing applications [28]. This research investigates the preparation of a water-soluble chitin derivative rather than chitosan because there are many studies on the preparation of chitosan water-soluble derivatives.

The solubility of chitin is achieved by the destruction of the strong hydrogen bonds in chitin molecules and their reorganization form a chitin gel [29]. Various procedures to make chitin water-soluble and its characterization are reported: [17, 30–53]. The water-solubility of chitin can be obtained by structural modification and by controlling the degree of deacetylation. Such modifications affect the properties of water-soluble chitin derivatives.

The degree of deacetylation of chitin/ chitosan plays a major role in their solubility in water. Modified chitins having 50% degree of deacetylation become soluble in water [54–57]. Deacetylated chitin has 50% degree of deacetylation with tosyl, iodo, trimethylsilyl and glucosyl groups soluble in water as well as in organic solvents [58]. This research mainly investigates on preparing water-soluble chitin derivatives having the degree of deacetylation similar to that of natural chitin.

2.3 Characterization of chitin and its derivatives

Characterization of chitin derivatives is by solubility, crystallinity, viscosity, degree of deacetylation, molecular weight, mechanical, thermal and moisture retention properties and antimicrobial properties and is helpful to determine their suitability in specific applications.

2.3.1 Solubility and crystallinity

Solubility is an important parameter for the use of chitin and its derivatives in a wide range of industrial applications. The sorption ability of chitin increases as the number of amino groups grows as the degree of deacetylation increases chitin is being converted into chitosan. The solubility characteristics of chitin/ chitosan are governed mostly by the extent of degree of deacetylation, the distribution of acetyl groups, degree of dissociation, processing methods, pH and the ionic strength [28, 59, 60]. The solubility of chitin and its derivatives can be demonstrated using chitosan in a dilute acidic medium. In this system, chitosan tends to be at equilibrium Eq. (1).

$$\text{Chitosan} - NH_2 + HOH \rightleftharpoons \text{Chitosan} - NH_3^+ + OH^- \quad (1)$$

Chitosan is soluble when pH is lower than 6 or 5.5. At lower pH, the amino groups in chitosan are fully protonated and the positively charged polymer chains will repel each other and fall apart in solution thus resulting in its dissolution. At pH above 6.5, chitosan will precipitate [54].

Polymer swelling reduces the crystallinity of chitin derivatives in a solution [61]. The crystallinity of chitin derivatives varies with the substitution of other functional groups onto the polymer chains [54]. The crystallinity index of chitin (85%) is higher than that of the water-soluble chitin derivative (48 to 57%) [42]. Because the acid or alkali treatments depolymerize the polymer chains during processing [62, 63]. Chitin shows a crystalline structure whereas water-soluble chitin derivatives show an amorphous structure due to structural modification [32, 54, 64].

2.3.2 Viscosity

Viscosity refers to the resistance to flow in liquids while elasticity refers to energy recovery in solids. Polymeric material may be time-dependent, acting more like a solid during short processing time (rapid movement) or acting more like a fluid during long processing time (slow movement). When a polymeric material has both fluid as well as solid behaviours, it is called viscoelastic. Like solubility, viscosity of chitin derivatives is also an important property in processing these polymers. The viscosity data also provides the information on the structure and properties of chitin-derived polymers [47]. This data will be helpful while designing new and innovative chitin-based films for various industrial applications. The dibutyryl chitin with an intrinsic viscosity of greater than 1 dl/g had a good spinnability and film-forming ability [65]. Apparent viscosity [η] or just viscosity, commonly used in place of dynamic viscosity, is defined as the ratio of the imposed shear stress [τ] to the shear rate [$\dot{\gamma}$]. See Eq. (2).

$$\eta = \tau/\dot{\gamma} \quad (2)$$

In a Newtonian fluid, the shear stress is proportional to shear rate, viscosity is therefore constant. If the viscosity of a fluid varies with respect to shear rate or shear stress, then it is termed a non-Newtonian fluid. A liquid such as water, alcohol etc, which is composed of a single substance is usually a Newtonian fluid. On the other hand, a polymer solution or a colloidal solution containing high molecular weight compounds and/or suspended solids is generally a non-Newtonian fluid.

Mathematical equations have been derived by many researchers to form flow models for describing the rheological behaviour of a material in terms of shear rate and shear stress. The simplest model that is used to describe the flow behaviour of a Newtonian fluid is Eq. (3). In case of non-Newtonian fluid, a power law model is usually used to describe the flow characteristics Eqs. (3) and (4) [66].

$$\tau = \eta\, \dot{\gamma} \tag{3}$$

$$\eta = K\, [\dot{\gamma}]^{n-1} \tag{4}$$

$$\tau = K\, [\dot{\gamma}]^{n} \tag{5}$$

where,

τ—Shear stress (Pa)

$\dot{\gamma}$—Shear rate (s^{-1})

η—Apparent viscosity (Pa. s)

K—Consistency coefficient (Pa. s^n)

n—Flow behaviour index, dimensionless number ($n = 1$ for Newtonian fluids; $n < 1$ for shear thinning or pseudoplastic fluids and $n > 1$ for shear thickening fluids).

Rheological study of both chitin, as well as water-soluble O-carboxymethylated chitin derivatives in N,N-dimethyl acetamide/ lithium chloride solvent system, exhibits non-Newtonian shear-thinning behaviour. The power law indices of these solutions increase along with the temperature [67]. Therefore, a solution containing chitin and/or its derivatives is a non-Newtonian fluid.

2.3.3 Degree of deacetylation (%)

The degree of deacetylation has an influence on all physiochemical properties such as molecular weight, viscosity and solubility. The presence of 50% amine groups defines the boundary between chitin and chitosan; chitin has less than 50% deacetylation and chitosan has more [68]. The degree of deacetylation is affected by the concentration of alkali, processing temperature, reaction time, previous treatment of chitin, particle size and chitin concentration. The degree of deacetylation and the distribution of the acetyl groups influence the solubility [69]. This research used the colloidal titration method to determine the degree of deacetylation of chitin and its derivatives [27, 32, 70].

2.3.4 Thermal analysis

A 'glass' can be defined as a solid, brittle material that has an amorphous liquid-like structure with very little flexibility or any obvious fluidity. Glass can be achieved by melting the ordered form of the material and then by rapid cooling or supercooling. A perfectly crystalline polymer will melt at a well-defined temperature; this melting transition is defined as a first-order transition. Melting causes discontinuous changes in volume, enthalpy and primary thermodynamic variables [71]. In an amorphous polymer, the molecular motion of the polymer chains is immobile at low temperatures. The state of the polymer is glassy. When the polymer is heated, the molecules obtain sufficient energy to slide over one another. The polymer becomes viscous, flexible or rubbery at the glass transition [72]. The glass transition temperature is highly specific to each anhydrous amorphous material and depends on experimental conditions, moisture content and molecular weight [73, 74].

The final glass transition temperature has significant impact on the final texture, diffusivity and the rate of deterioration. Stiffness or brittleness of the polymers is lost by the reduction in the glass transition temperature owing to the plasticizing effect of water. This will make the polymer unsuitable for making films [74–76]. Thermal behaviour of chitin derivatives is conducted by thermomechanical analysis (TMA), thermogravimetric analysis (TGA) and differential scanning calorimetry (DSC). DSC is an effective technique to evaluate the thermal behaviour as well as to determine the degree of deacetylation of chitin derivatives [64, 77].

Figure 1 shows the typical chromatogram of a chitin derivative during DSC analysis. In the DSC chromatogram, the first endothermic peak indicates the loss of absorbed moisture by the films. The second exothermic peak(s) indicates the degradation of chitin derivatives. Finally, the phase transition of the chitin polymer occurs. Various studies have been performed to analyze chitin and its derivatives using DSC [58, 77–80]. Both chitin and water-soluble carboxymethyl-chitin exhibit the endothermic peak that relates to the loss of water during DSC [64]. The decomposition of chitin during DSC is obtained in two stages: one peak at a temperature range of 200 to 260°C and the second peak at 300°C to 360°C [79]. In contrast, a single-stage decomposition of chitin is obtained at around 400°C [77]. Chitin shows better thermal stability than chitosan it contains fewer amine groups [77]. The increase in molecular weight causes a proportional increase in glass transition temperature. DSC analysis of

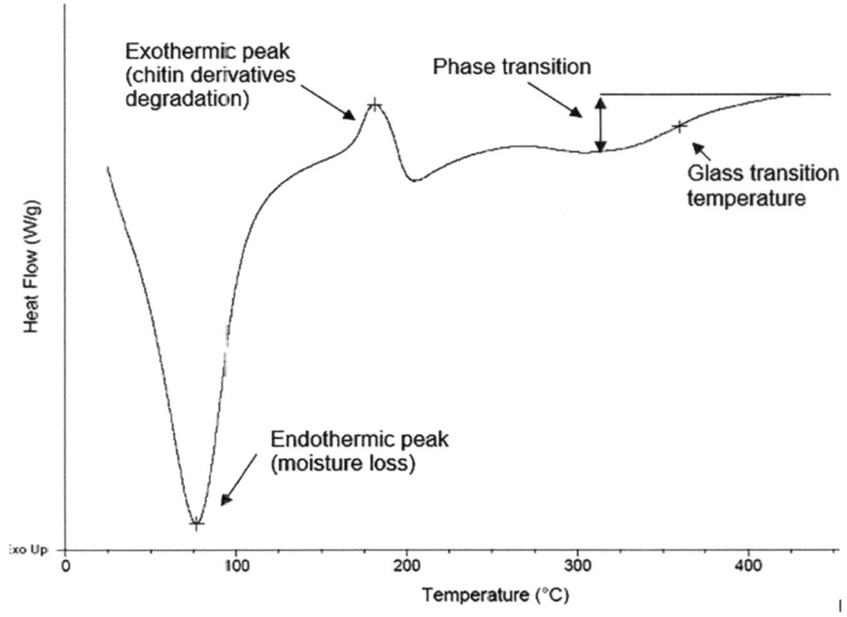

Figure 1.
DSC chromatogram of a chitin derivative.

high molecular weight chitin and chitosan shows no glass transition even up to a temperature of 550°C [64, 77, 79].

2.3.5 Moisture absorption

Food packaging technology requires the use of low oxygen and carbon dioxide permeable materials. The presence of water in the polymer influences the way in which these gases are sorbed and diffused [81]. Biomedical or pharmaceutical activity depends on how the water molecules are associated with the polymer. Moreover, the swelling characteristics of polymer gels are dominated by the nature of the polymer and the state of water [82]. Moisture absorption in polymeric films is important for a variety of industries ranging from microelectronics to adhesives and coatings. In many applications, water absorption leads to reliability problems such as the degradation of dielectric properties, corrosion or delamination. A significant number of studies covering many polymer systems have focused on characterizing the absorption and diffusion properties of water in polymer films [83]. It is very interesting to obtain a better understanding of the water sorption phenomena and the mechanical strength of chitin-based films prior to their use in food, medical and pharmaceutical applications. The solubility and strong swelling of the finished films in water decrease the stiffness of the films [47].

Better understanding of moisture absorption mechanisms and controlling steps may help not only in optimizing the use of chitin films but also in designing new chitin-based polymers. Equilibrium moisture absorption properties are frequently controlled by diffusion properties [especially intra-particle diffusion], degree of deacetylation, chemical structure and physical modification of the polymer; when the size of sorbent particles increases, moisture absorption performance may drastically decrease, the time required to reach equilibrium exponentially increases and sometimes the sorption capacity at equilibrium diminishes [64, 81, 84]. The moisture absorption ability of water-soluble chitin derivatives depends on their chain conformation in solution and their molecular weight [39].

2.3.6 Antimicrobial properties of chitin

Although there are many studies about the antimicrobial activity of chitosan, few studies have been performed to analyze the antimicrobial activity of natural chitin or water-soluble chitin. This is because chitosan has higher degree of deacetylation. An increase in the degree of deacetylation of chitin and hence the number of amino groups increases the antimicrobial activity [26, 85–87]. Therefore, there is a relationship between the antimicrobial activity and the degree of deacetylation of chitin and its derivatives. This research mainly focuses on the antibacterial property of water-soluble chitin derivatives against *Bacillus cereus*.

3. Materials and methods

3.1 Materials used

Due to the difficulty in sourcing adequate commercial prawn waste, raw eastern school prawns (*Metapenaeus macleayi*, approximately 9 cm body length) were obtained from Department of Primary Industries, Fisheries Conservation Technology

Unit, NSW, Australia and hand-peeled to obtain prawn waste to conduct this research (moisture content 74%; Ash 23% and 11% chitin dry basis). This prawn waste was stored at –22°C until used for research. All solvents used were HPLC grade supplied by Lab Scan Analytical Sciences. All chemicals were AR grade.

3.2 Preparation of water-soluble colloidal chitin (WSCC)

Step 1: Natural chitin was recovered from prawn waste before it was converted into WSCC. First, prawn pigment, astaxanthin complex was extracted from prawn waste followed by deproteinization using 10% sodium hydroxide (1:2, w/v) at 100°C for 6 hours and then demineralization using 2M hydrochloric acid (1:3, w/v) for 48 hours at ambient temperature. The residue was natural chitin [88]. The moisture content chitin and ash content were measured using Equation 2.8 [89]. Care was taken that the ash content of chitin was less than 1%.

Step 2: This natural chitin was freeze-dried at ambient temperature for 24 hours at 0.4 mbar and stored in the desiccator. Finely ground freeze-dried natural chitin (100 to 125 mesh) (5%, w/v) was digested in sodium hydroxide (50%, w/v) for 6 hours at 87°C to prepare chitosan. The chitosan was thoroughly washed to neutral pH and freeze-dried [88].

Step 3: Chitosan hydrochloride was prepared by dissolving 2.5 g of freeze-dried chitosan in 100 mL of 10% acetic acid followed by precipitation using concentrated hydrochloric acid. The precipitate was thoroughly washed using methanol several times to get chitosan hydrochloride free of chloride. The presence of Cl^- was tested by adding 1% silver nitrate to the filtrate; a white precipitate indicated presence of Cl^-. The chloride-free precipitate called chitosan hydrochloride was dried in the oven at 50°C and moisture content was determined.

Step 4: Two methods were studied to prepare WSCC.

Method 1: Oven-dried chitosan hydrochloride (0.5 g) was dissolved in 2 mL of distilled water and then re-acetylated by adding a mixture of acetic anhydride (5 mL) and pyridine (2.5 mL). The reaction mixture was stirred in a magnetic stirrer overnight at ambient temperature to evaporate the solvent. This WSCC was dried in the oven at 50°C and the moisture content was measured.

Method 2: Oven-dried chitosan hydrochloride (0.5 g) was re-acetylated by mixing with an equal volume of acetone (5 mL) and acetic anhydride (5 mL). The mixture was stirred in a magnetic stirrer overnight at ambient temperature to evaporate the solvent. It was then dried in the oven at 50°C. The moisture content of the oven-dried WSCC was measured.

3.3 Preparation of WSCC film for antibacterial study

WSCC film of 0.22 mm thickness was prepared by pouring 15 mL of WSCC solution in water (5%, w/v) on a glass petri dish underlined with a layer of microwave-safe all-purpose food packaging film without imperfections. WSCC film of 0.52 mm thickness was prepared by pouring 100 mL of WSCC solution in water (5%, w/v) on a glass plate (16 cm × 24 cm) underlined with a layer of GLAD wrap. These two samples were then oven-dried at 40°C. Careful attention was given that the prepared film was free from air bubbles and physical imperfections. The thickness of the dry WSCC films was measured using a vernier calliper at five random positions and averaged.

3.4 Characterization of WSCC

3.4.1 Viscosity of WSCC using different viscometers

In this study, the viscosity of WSCC dissolved in distilled water was evaluated using two viscometers. They are rotational Haake viscotester® VT550 and SV-10 AND Vibro viscometer (SV series 300). The results obtained from these two viscometers were then compared. Rotational Haake viscometer measures the viscosity by measuring the running torque of the cylindrical rotors immersed in a sample because viscosity is directly proportional to a running torque required to develop steady rotating motion. Its temperature is controlled by a re-circulating water bath and water jacket. SV-10 AND Vibro viscometer measures the viscosity by controlling the amplitude of the sensor plates immersed in a sample and measuring the electric current to drive the sensor plates. SV-10 AND viscometer vibrates with sine-wave of frequency about 30 Hz and amplitude of approximately 0.2 mm. The temperature at the geometric centre is measured, not controlled. These two viscometers have different shear rates. The shear rate of the Haake viscometer is controlled. The nominal shear rate values for SV-10 AND Vibro viscometer are shown in **Table 1**.

3.4.2 Viscosity testing using Haake viscometer

In the Haake viscometer, a sample solution (9 mL) of WSCC (5%, w/v) in distilled water was taken in a concentric cylinder (NV type, system no: 8, radius: 20.5 mm) together with an NV rotor (system no: 8, radius: 17.85 mm, height: 60 mm). Prior to the analysis, the sample solution and sensor system were placed in the water bath at the respective temperature for 15 minutes for equilibration. The flow curves were observed at shear rates 0.13 to 300 s^{-1} and at temperatures 20°C, 30°C and 40°C. The parameters for power-law models were noted. The operating parameters used during viscosity analysis are given below. Ramp1 eliminated start-up effects and was not taken into account while analyzing the results.

Ramp1: CR 0.13 s^{-1}; t = 10 s; #5
Ramp2: CR lin, 0.13–300 s^{-1}; t = 120 s; #100
Ramp3: CR lin, 300–0.13 s^{-1}; t = 120 s; #100

Power-law models of flow curves were calculated for each temperature Eq. (4). The apparent viscosity at shear rates 5, 10, 100 s^{-1} was calculated from each model Eq. (5) [66]. The temperature dependence of viscosity follows the Arrhenius exponential relationship Eq. (6).

Viscosity coefficient (mPa.s)	Maximum shearing rate (s^{-1})	Effective shearing rate (s^{-1})
1	590	420
10	130	92
100	42	30
1000	17	12
10000	10	7

Table 1.
Nominal shear rate values of a viscosity standard Newtonian fluid using SV-10 AND Vibro viscometer.

$$\eta = Ae^{E/RT} \tag{6}$$

where,
E—Activation energy of the sample
A—Empirical constant
T—Temperature of reaction mixture
R—Universal gas constant
η—Apparent viscosity

This exponential relationship of Arrhenius applies to a polymer system of low molecular weight and low viscosity [90]. From the results obtained, WSCC was a low molecular weight polymer with low viscosity. Therefore, Arrhenius models were constructed for WSCC solutions using Eq. (7). Arrhenius models were obtained by plotting the natural logarithmic of apparent viscosity (η) on Y-axis and the inverse of temperature (K^{-1}) on X-axis for each shear rate. Arrhenius constants were determined from the slope and the intercept of the Arrhenius model Eq. (7). Activation energy values (E) at each shear rate were calculated by multiplying the respective slope of the Arrhenius equation with the gas constant value (R). The intercept was the natural logarithmic of the empirical constant, A.

3.4.3 Viscosity testing using SV-10 AND viscometer

Sample solution (10 mL) of WSCC (5%, w/v) in distilled water was analyzed using SV-10 AND Vibro viscometer, which delivers a single shear rate, and the temperature and viscosity profile were recorded over time. An Arrhenius plot was constructed and Arrhenius constants were determined.

Apparent viscosity values (η_S) at 20°C, 30°C and 40°C were interpolated using the Arrhenius model, and the corresponding shear rates ($\dot{\gamma}$) were calculated from the effective shear rate *vs* viscosity co-efficient standard curve. The calculated shear rates were then substituted in the power-law equations of Haake viscotester Eqs. (4) and (5). The apparent viscosity values (η_H) corresponding to the SV-10 AND vibratory viscometer shear rates at 20°C, 30°C and 40°C were calculated. Finally, the apparent viscosities of WSCC in distilled water were obtained using both the viscometers and were compared.

3.4.4 Degree of deacetylation determination

Degree of deacetylation of chitin samples was measured by the colloidal titration method. Oven-dried chitosan hydrochloride (1 g) samples prepared from methods 1 and 2 were titrated against 0.1M sodium hydroxide using phenolphthalein indicator. The degree of deacetylation of the sample was calculated Eq. (7) [91]. Standard deviation and significant differences were calculated.

$$\%DD = \frac{N1V1}{1000} x \frac{V0}{V2} x(MWCTS - Cl)x \left\{ \frac{100}{W4x(1 - \%MC/100)} \right\} \tag{7}$$

where,
%DD—Degree of deacetylation of chitin sample (%)
N_1—Concentration of sodium hydroxide (M)
V_1—Volume of sodium hydroxide used (mL)

V_0—Total volume of chitosan chloride solution (mL)
V_2—Volume of chitosan chloride solution used for titration (mL)
MW_{CTS-Cl}—Monomer weight of chitosan chloride
W_4—Weight of chitosan chloride taken for titration (g)
%MC—Moisture content of chitosan chloride (%)

3.4.5 Differential Scanning Calorimetric [DSC] analysis of WSCC film

Differential scanning calorimetric measurements were performed using Universal V4.3A TA Instruments. WSCC films [sample weight 3 mg, 0.22 mm thickness] were equilibrated in a desiccator or in the relative humidity chamber (92.5%) for a week and the DSC analysis was performed under a dynamic nitrogen atmosphere (50 mL/minute) at a heating rate of 5°C/minute. Samples equilibrated to a range of relative vapour pressures (36.1%) between these two were also tested. Intermediate relative humidity (36.1%) of WSCC film was obtained by equilibrating the film at ambient conditions. Accurately weighed sample ($\pm$0.1 mg) was placed into a covered aluminium sample holder. An empty sample holder was used as reference and two runs were performed for each sample by heating the sample from 25°C up to 450°C.

In another study, two samples treated at the ambient relative humidity were tested individually using DSC. The DSC curves were performed under a dynamic nitrogen atmosphere (50 mL/minute) at a heating rate of 5°C/minute. Accurately weighed samples ($\pm$0.1 mg) were placed into an aluminium sample holder and sealed. An empty sample holder was used as reference and the runs were performed by heating the samples from 25°C up to 110°C with an isothermal for 15 minutes to remove the moisture present in the samples. The samples were then reweighed and reheated from 25°C up to 480°C [77]. The results are then compared with the phase behaviour of the moist WSCC films.

3.4.6 Moisture absorption isotherm of WSCC film

In this study, moisture absorption behaviour of WSCC was investigated at different relative humidities. Moisture absorption isotherm of WSCC film of 0.22 mm thickness was prepared using saturated salt solutions of different relative humidities (**Table 2**). WSCC films (2.5 × 2.0 cm) were placed in each of the relative humidity chambers and the samples were kept at a controlled temperature of 25°C for equilibration. The moisture gained by each of the samples was measured after a week and

No.	Saturated Solutions	Equilibrium relative humidity (%) at 25°C
1	Potassium acetate [CH_3COOK]	22.5
2	Magnesium chloride [$MgCl_2$. $6H_2O$]	32.7
3	Potassium carbonate [K_2CO_3. $2H_2O$]	43
4	Sodium nitrite [$NaNO_2$]	64
5	Sodium chloride [NaCl]	75.1
6	Potassium chloride [KCl]	84.2
7	Potassium nitrate [KNO_3]	92.5

Table 2.
Relative humidity standards used for the experiment [83, 92].

the moisture absorption isotherm was prepared by plotting the relative humidity on X-axis and the moisture content of the film on Y-axis.

3.4.7 Antibacterial activity of WSCC film

Bacillus cereus 043800 was supplied by the culture collection of Department of Biological Sciences, UNSW and stored at –80°C freezer. The inoculum suspension (10 μL) was spread on WSCC film (2 cm × 1.5 cm). A control WSCC film was made without inoculum. Each treatment was carried out in duplicate. The films were then placed in the aseptic plastic Petri plates and autoclaved glass Petri plates separately. Petri dishes containing the films were then sealed using adhesive tape and incubated at 30°C for a day. Then inoculated films were taken out and the films were placed into a 'stomacher' bag with 10 mL of peptone/water (0.1%, w/v). The stomacher bags were initially massaged by hand to loosen the adhesion of cells to the film followed by 'stomaching' using stomacher for 5 minutes and then allowed to sit for 5 to 10 minutes. The bag was again massaged by hand before plating to ensure homogeneous distribution of the suspension. 50 μL of each of these samples was spread-plated onto a brain heart infusion agar plates. The plates were incubated at 30°C, and colonies were counted 1 day later [93].

4. Results and discussion

4.1 Solubility of WSCC

The WSCC prepared by Method 1 was soluble in water and the oven-dried material was not readily soluble in water. Moreover, this method was not reproducible, and the strong odour of pyridine was other major concern. When the processing conditions of chitin to chitosan were changed in Method 2, the degree of deacetylation of the prepared chitosan was less and the oven-dried WSCC readily formed a colloidal suspension in water. Method 2 was reproducible in preparing WSCC and this method did not use pyridine. Therefore, Method 2 was chosen to produce WSCC from prawn waste.

4.2 Characterization of WSCC

4.2.1 Viscosity of WSCC using different viscometers

4.2.1.1 Viscosity testing using Haake viscometer

Flow curves obtained for WSCC dissolved in distilled water (5%, w/v) using Haake viscometer at different temperatures and different shear rates were modelled by power-law equations. The value of the power law exponent, n was less than 1 (ranging between 0.8 and 0.9) at all temperatures Thus, WSCC dissolved in distilled water was a shear-thinning, non-Newtonian fluid. As would be expected, the apparent viscosities generally decreased with an increase in temperature and shear rate during the experiment. A flow curve obtained for WSCC dissolved in distilled water (5%, w/v) at 30°C is shown in **Figure 2**.

The Arrhenius model for WSCC in distilled water using Haake viscometer at different shear rates and at different temperatures is shown in **Figure 3**. Arrhenius

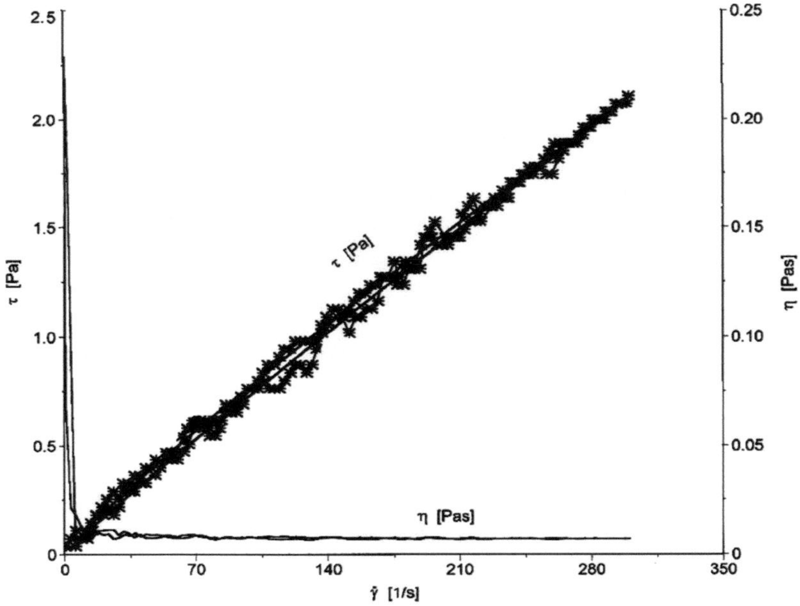

Figure 2.
Flow curve obtained for WSCC dissolved in distilled water (5%, w/v) at 30°C using Haake viscometer.

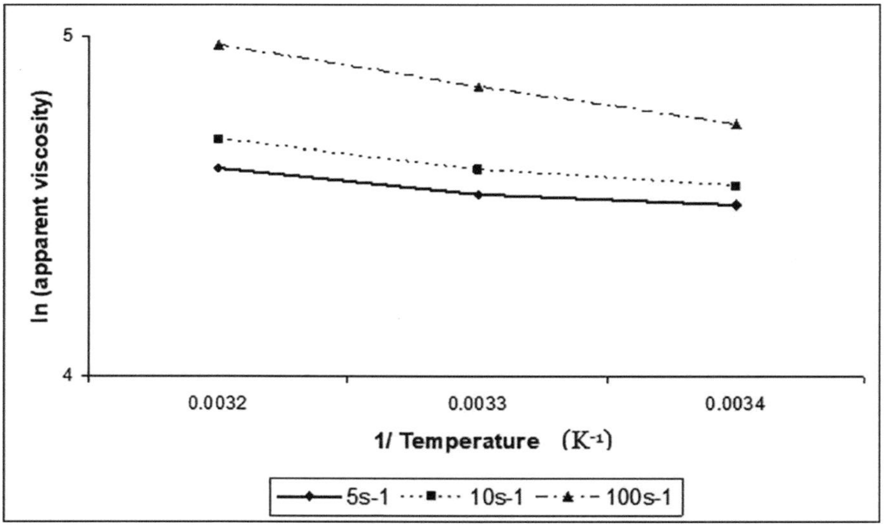

Figure 3.
Arrhenius models of WSCC in distilled water (5%, w/v) at different shear rates and at different temperatures: Haake viscometer.

models at different shear rates were straight lines with negative slope, hence fit into Arrhenius exponential relationship. The R^2 values were between 0.94 and 1.00. An increase in temperature decreased the apparent viscosities at different shear rates (5 s^{-1}, 10 s^{-1} and 100 s^{-1}). Therefore, temperature dependence of apparent viscosity of WSCC in distilled water obeyed the Arrhenius exponential relationship. The

Arrhenius model constants at different shear rates for this particular sample system are shown in **Table 3**. In this sample system, the activation energy values increased with an increase in temperature (positive activation energy values) and shear rate.

4.2.1.2 Viscosity testing using SV-10 AND vibratory viscometer

The relationship between time, temperature and apparent viscosity profile of WSCC in distilled water (5%, w/v) using SV-10 AND vibratory viscometer is shown in **Figure 4**. As expected, the apparent viscosity decreased with increase in temperature.

The Arrhenius model of natural logarithmic values of apparent viscosity against the reciprocal of absolute temperature for WSCC in distilled water (5%, w/v) using SV-10 AND viscometer is shown in **Figure 5**. The relationship was linear with negative slope (R^2 = 0.99), thus followed the Arrhenius exponential relationship. The Arrhenius model for this system obtained using SV-10 AND viscometer was y = −8252.4x + 32.297. The intercept, which is the natural logarithmic value of the empirical constant, was –32.297. The activation energy for this sample system calculated

Shear rate (s^{-1})	Activation energy, E (KJ/mol.K)	Empirical constant, ln(A)
5	0.0005	−4.6568
10	0.0006	−4.7567
100	0.001	−5.0889

Table 3.
Arrhenius model constants of WSCC in distilled water (5%, w/v) at different shear rates: Haake viscometer.

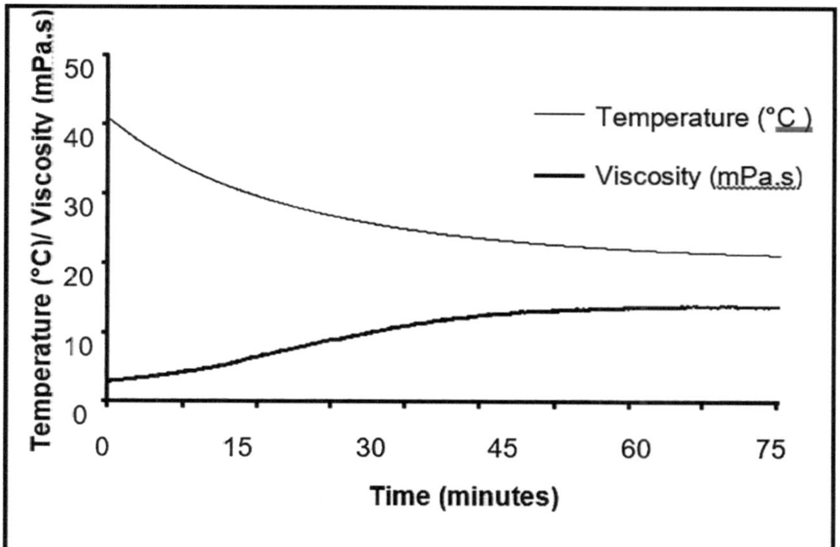

Figure 4.
Temperature and apparent viscosity profile of WSCC in distilled water (5%, w/v) using SV-10 AND vibratory viscometer.

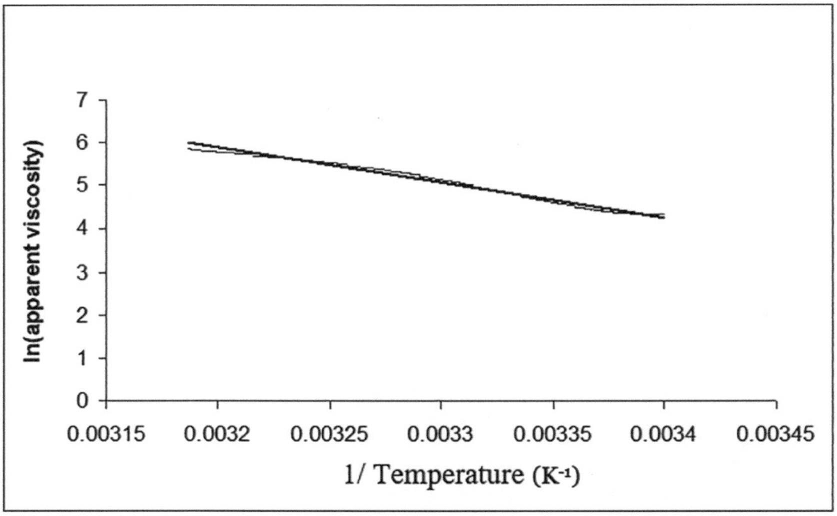

Figure 5.
Arrhenius model of WSCC in distilled water (5%, w/v) at a single shear rate: SV-10 AND vibratory viscometer.

from the slope of the Arrhenius model was 68.61 KJ/mol.K, which is a positive value. Positive activation energy for the same sample system was obtained in the Haake viscometer as well (**Table 3**).

The Arrhenius models constructed for WSCC in distilled water (5%, w/v) using both viscometers showed negative slope. The Arrhenius models obtained from both viscometers confirmed that the apparent viscosity decreased with an increase in temperature. As mentioned in *3.4.1*, the apparent viscosities (η_S) of WSCC dissolved in distilled water (5%, w/v) at 20°C, 30°C and 40°C were calculated from the Arrhenius model of the SV-10 AND vibratory viscometer, which are shown in **Table 4**. The corresponding shear rate values for the calculated apparent viscosities at 20°C, 30°C and 40°C were obtained from the effective shear rate *vs* viscosity standard curve of the SV-10 AND vibratory viscometer. These values are listed in **Table 4**. The apparent viscosities were not constant with varying shear rates. Thus, WSCC in distilled water (5%, w/v) behaved like a non-Newtonian fluid. A similar result was obtained when analysed using the Haake viscometer.

The obtained shear rate values from the SV-10 AND vibratory viscometer values at 20°C, 30°C and 40°C were then substituted into the power-law model of Haake viscometer. The apparent viscosities (η_H) for those shear rates were then worked out using Haake viscometer power-law equations. The obtained apparent viscosities (η_H) are given in **Table 4**.

Temperature (°C)	Apparent viscosity (mPa.s)	Shear rate (s^{-1})	Apparent viscosity η_H (mPa.s)
20	15.82	48.55	9.43
30	6.25	32.14	8.75
40	2.62	21.85	8.32

Table 4.
Apparent viscosities and shear rates of WSCC in distilled water (5%, w/v) at different temperatures: SV-10 AND vibratory viscometer.

The apparent viscosities for the respective shear rates obtained using both viscometers were different (**Table 4**). The differences in the calculated apparent viscosities by these two viscometers are due to the different working principles. In Haake viscometer, shear rate can be controlled whereas the shear rate cannot be controlled in SV-10 AND Vibro viscometer. Both these viscometers showed that the system containing WSCC in distilled water (5%, w/v) was a non-Newtonian fluid with shear thinning behaviour and an increase in temperature reduced the apparent viscosities. Similar results are obtained for chitin and water-soluble *o*-carboxymethylated chitin dissolved in *N*,*N*-dimethyl acetamide/ lithium chloride solvent system [67]. Thus, the apparent viscosity of WSCC dissolved in distilled water is dependent on the temperature and the shear rate.

4.2.2 Degree of deacetylation of WSCC

There was no significant difference between the degree of deacetylation of chitin and WSCC (**Table 5**). This is because acetylation with a mixture of acetic anhydride-acetone gives rise to the complete acetylation of amino groups [94, 95]. Thereby the original degree of deacetylation of chitin was restored by WSCC. The yield of WSCC was 82 g/kg of prawn waste (dry basis).

4.2.3 Differential Scanning Calorimetric (DSC) analysis of WSCC film

The DSC curves obtained for WSCC films equilibrated in high relative humidity (92.5%) (moisture: 16.4%) and in a desiccator (moisture: 6.52%) are shown in **Figures 6** and 7 with exothermic peaks facing up. A generic DSC curve obtained for the WSCC film equilibrated in high relative humidity is also shown in **Figure 1**. Samples equilibrated to a range of relative vapour pressures between these two were also tested. The DSC thermogram for both these samples showed an endothermic peak (Peak 1) after 70°C followed by an exothermic peak around 180°C (Peak 2).

Various studies indicate that the endothermic peak at 100°C is attributed to the evaporation of absorbed water and the first exothermic peak is probably due to the degradation of chitosan studied during DSC [64, 82, 96–98]. In another study, the endothermic peak corresponds to the loss of moisture seen at a lower temperature (70°C), which is similar to this experiment [77]. The onset of the endothermic peak is related to pressure build-up because of water evaporation inside the sealed sample cups during DSC. The pressure at which the seam of the cups started to leak corresponds to approximate vapour pressure of water. The leaking of sealed sample pans during DSC also relates to the sample weight loss after the run [64].

Peak 1 in this study is due to loss of absorbed moisture by the WSCC film. As expected, a smaller peak of water loss (Peak 1) for a low moist sample was obtained

Sample	Degree of deacetylation (%)
Chitin	27.64 ± 0.61
Chitosan (Method 1)	88.22 ± 0.46
Chitosan (Method 2)	54.08 ± 0.69
WSCC	28.46 ± 1.08

Table 5.
Degree of deacetylation of chitin and its derivatives.

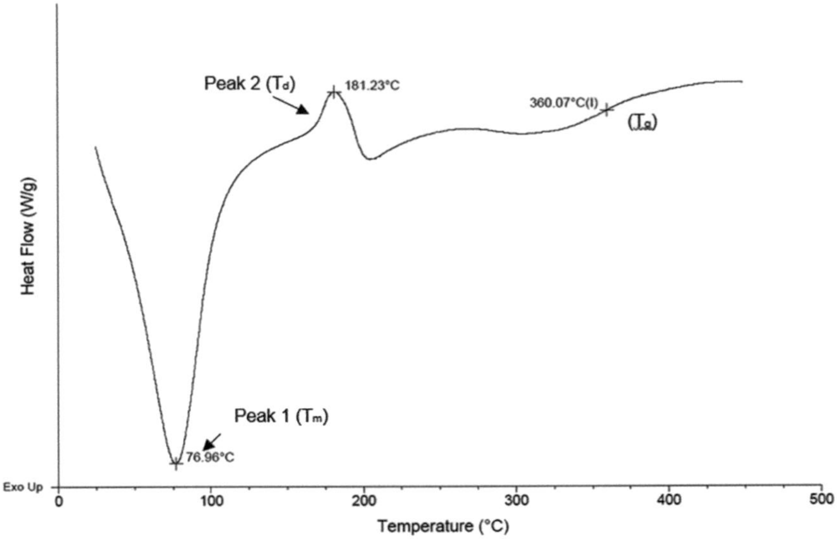

Figure 6.
DSC curve for high-moisture WSCC film.

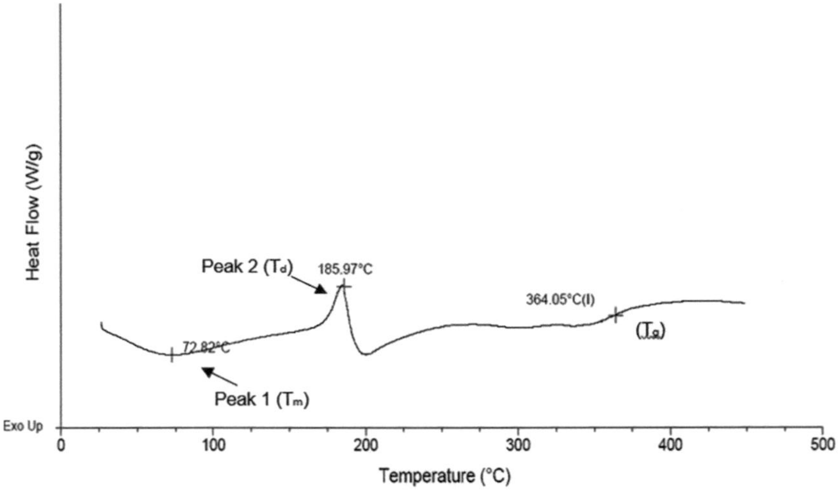

Figure 7.
DSC curve for dry WSCC film.

compared to a high moist sample. It was also expected that the sample equilibrated in the desiccator would be truly anhydrous. However, sample treated in desiccator showed endothermic peak for water loss. Some water in the film was not completely removed during sample drying in the desiccator. Therefore, it is suggested that desiccators be evacuated to accomplish faster and complete removal of water. Moreover, totally anhydrous chitin samples are difficult to obtain because of the high water affinity presented by these polymers [77].

In this study, Peak 2 is due to the degradation of chitin derivatives. The first exothermic peak for degradation of chitin derivatives is obtained at around 180°C in this study. The exothermic peak for the decomposition of chitin derivatives depends on the molecular weight and the presence of hydrophilic groups [64]. Viscometric analysis of WSCC shows that the molecular weight of WSCC is lower than chitin. Because of this reason, low molecular weight WSCC derivatives degrade faster than the high molecular weight chitin. The degradation of WSCC film occurred at relatively lower temperature than the published data of chitin and water-soluble carboxymethyl-chitin run [64]. The thermal stability of WSCC is poorer compared to chitin and water-soluble carboxymethyl-chitin.

The degradation temperature (T_d) of high-moist WSCC film was low compared to low-moist WSCC film. However, noticeable differences in the degradation temperature of both low-moisture and high-moisture samples were hard to see (**Table 6**). WSCC film did not show a considerable increase in the absorption of moisture over a wide range of relative humidities (**Figure 8**). Therefore, moisture content of WSCC film slightly influences the degradation of WSCC derivatives in this study.

The glass transition temperature (T_g) for WSCC film was observed at around 360° C. The glass transition temperature (T_g) of high-moisture WSCC film was lower than that of low-moisture WSCC film (**Table 6**). Higher molecular mobility accelerates reactions limited by diffusion and decreases stiffness [84]. The thermal degradation of the material is sensitive to moisture content confirms that degradation reactions were

Samples	**T_m (°C)**		**T_d (°C)**		**T_g (°C)**	
	Run 1	**Run 2**	**Run 1**	**Run 2**	**Run 1**	**Run 2**
Low-moisture WSCC film	72.82	70.32	185.97	186.91	364.05	364.45
Standard Deviation	1.41		0.67		0.28	
High-moisture WSCC film	76.96	77.69	181.23	185.10	360.07	363.48
Standard Deviation	0.52		2.74		2.41	

Table 6.
Glass transition temperature (T_g) of WSCC films determined using DSC.

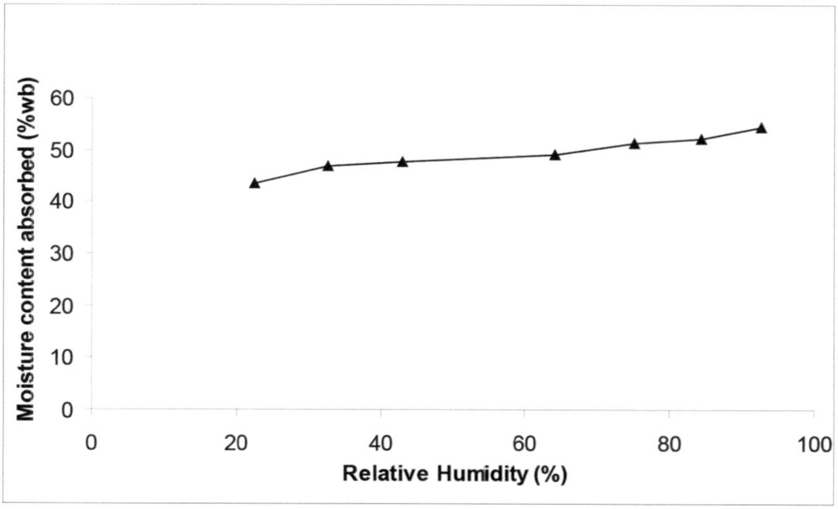

Figure 8.
Moisture absorption isotherm of WSCC film.

diffusion-limited. Therefore, moisture content did influence the phase behaviour of WSCC film. There was, however, sample weight loss in both high-moisture (57%) and low-moisture (50%) WSCC films after DSC run. The difference in the sample weight loss is due to the difference in the moisture content. The sample weight-loss (15%) of water-soluble carboxymethyl-chitin film after DSC run is also reported by Ref. [64]. It was beyond the scope of this project to exhaustively test the rheological and thermophysical properties of chitin and chitosan films against a range of other polymers. Such tests would be more profitably carried out at a larger scale of production, which would enable rolled or extruded films to be made. Nevertheless, the tests done show that the recovery of prawn chitin and its processing can lead to films of high and reproducible quality.

4.2.4 Moisture absorption isotherm of WSCC film

The specific moisture absorption isotherm of WSCC film (0.22 mm thickness) appeared to follow the expected sigmoid curve usually obtained over the whole range (0 to 1) of relative vapour pressure (**Figure 8**). This kind of isotherm is also called a type II isotherm in reference to the Brunauer–Emmett–Teller (BET) model [99]. A similar result was obtained for chitosan films by Ref. [99].

The specific moisture absorption of WSCC film increased with increasing relative humidity. The absorbed moisture by WSCC films varied from 40% to 55% between low (22%) and high relative humidity (92.5%). In this case, there was not much difference obtained in terms of the absorption of specific moisture over a wide range of relative humidities. The other study shows that the absorbed moisture content by chitosan films varies from 5 to 45% over relative humidities from 20% to 80% [99]. This indicated the moisture absorption of WSCC films was less compared to chitosan films over a wide range of relative humidities. The solubility and strong swelling of the finished film in water decreases the stiffness of the films which in turn decreases the suitability of the film in various applications [47]. The specific moisture absorption behaviour of WSCC film may not vary considerably over a range of relative humidities. This property will increase the selectivity of the finished films when they are used as membranes, packaging films or coating materials. In-depth study on the moisture absorption mechanism of WSCC films will be carried out during the scale-up of this process. Moisture absorption isotherm of WSCC films will be useful to determine the suitability of the film for maintaining proper moisture content of a particular product.

4.2.5 Antibacterial activity of WSCC film

The antimicrobial activity of WSCC film against the gram-positive bacterium *Bacillus cereus 043800* was investigated. The average values of the number of colonies of *B. cereus* during the antibacterial study are presented in **Table 7**. The empty glass and plastic Petri plates incubated alone for sterility check did not show the presence of *B. cereus* on brain heart infusion agar. WSCC film inhibited around 90% of the growth of *B. cereus 043800*. This shows that films made by WSCC have excellent antibacterial activity against *B. cereus*. The interesting finding in this study is that although the degree of deacetylation of WSCC was less (28.5%), WSCC film inhibited almost 90% of the growth of *Bacillus cereus 043800*. The extension of this work would test the overall antimicrobial activity of WSCC film against different microorganisms. Recently, water-soluble chitin derivation was investigated for the film coating of

Antibacterial study details	Plastic Petri plate	Glass Petri plate
(A) Cell concentration in *stock* suspension (cfu/mL)	8.7×10^7	8.7×10^7
(B) Cell concentration in *inoculum* suspension (cfu/mL)	2.98×10^6	2.63×10^6
(C) Average number of bacteria in the control (WSCC film without bacterial *inoculum*) (cfu)	1.98×10^4	2.13×10^4
(D) Average no. of bacteria inoculated onto film (cfu)	2.98×10^4	2.63×10^4
(E) Average no. of bacteria present in the film after incubation (cfu)	3×10^3	3.1×10^3
Bacterial colonies inhibited (100−(E/D × 100)) (%)	89.2	88.2

Table 7.
Concentration of Bacillus cereus 043800*: An antibacterial study of WSCC film.*

Ricotta cheese, and it proved to be efficient in prolonging the shelf life of Ricotta cheese [17].

5. Conclusions

In conclusion, this study has clearly demonstrated the fundamental characteristics of WSCC. The study of WSCC will be helpful in evaluating its suitability in various industrial sectors including food and pharmaceuticals. To effectively use WSCC as a functional ingredient, relationships between the functional properties and characteristics of WSCC must be constantly monitored for proper quality control. In the current study, limited relevant information on aspects of such relationships was obtained. More extensive investigations are needed for a better understanding of the relationships reported in the present research, especially in view of current worldwide interest in commercial use of water-soluble chitin derivatives. This will be best done during the scale-up of this current project.

The biodegradable water-soluble colloidal chitin recovered from prawn waste offers exciting possibilities because of its water solubility. These chitin derivatives can be obtained from the prawn waste changing the waste stream into a valuable resource that is commercially viable. The antibacterial property of water-soluble colloidal chitin may be applied in pharmaceutical, medical and food packaging and coating applications. Biodegradable films made from these natural polymers obtained from renewable sources will protect the environment in the future and will be an effective alternative to plastics that threatens the environment. However, vegetarians may object to the use of animal polymers.

6. Recommendations

The study of the preparation of water-soluble colloidal chitin from prawn waste recommends scale-up to pilot scale and reexamination of the stability of water-soluble colloidal chitin. Further microbial investigation is required to explore the water-soluble colloidal chitin film's antimicrobial activity on a wide range of microorganisms. An appropriate method needs to be developed to determine the molecular weight and physio–chemistry of the water-soluble colloidal chitin following scaled-up production. Detailed study about the toxicity of water-soluble colloidal chitin is necessary prior to its use in industrial applications, especially in food applications because these compounds may contain chemical/solvent residues. Water-soluble colloidal chitin can be tested as an antimicrobial coating in fruits and vegetables. The application of water-soluble colloidal chitin in food packaging applications can be studied in the future.

Further, the use of sodium hydroxide and hydrochloric acid during preparation of water-soluble colloidal chitin generates chemical waste. World over efforts is on to find out an alternative to chemical method of preparation of chitin/ chitosan. Therefore, an alternative way of preparing water-soluble chitin derivatives using enzymes or simply by adding or substituting functional group to chitin molecule is recommended. Such material can be used for preparing films and the related studies.

Acknowledgements

I acknowledge Prof. Wilem F. Stevens, Mahidol University, Thailand, who guided me during this research, when required. I also acknowledge Katherine Zerdin, CSIRO, Sydney, Australia for generously sharing the resources to perform the antibacterial work.

Notes/thanks/other declarations

Many thanks to the Department of Primary Industries, Fisheries Conservation Technology Unit, NSW, Australia for supplying raw Eastern School prawns for the research. I also thank the Australian Government for the financial support by providing the International Postgraduate Research Scholarship to undertake this research.

Author details

Renuka Vinothkumar* and Janet Paterson
University of New South Wales, Sydney, Australia

*Address all correspondence to: renuevino@yahoo.com

References

[1] Marsh K, Bugusu B. Food packaging—Roles, materials, and environmental issues. Journal of Food Science. 2007; **72**(3):R39-R55

[2] Srinivasa PC, Tharanathan RN. Chitin/chitosan—Safe, ecofriendly packaging materials with multiple potential uses. Food Reviews International. 2007;**23**(1):53-72

[3] Hirano S, Zhang M, Chung B. The N-acylation of chitosan fibre and the N-deacetylation of chitin fibre and chitin-cellulose blended fibre at a solid state. Carbohydrate Polymers. 2000;**41**: 175-179

[4] Hayes M, Carney B, Slater J, Bruck W. Mining marine shellfish wastes for bioactive molecules: Chitin and chitosan—Part A: Extraction methods. Biotechnology Journal. 2008;**3**(7):871-877

[5] Amar B, Philip R, Bright Singh IS. Efficacy of fermented prawn shell waste as a feed ingredient for Indian white prawn, *Fenneropenaeus indicus*. Aquaculture Nutrition. 2006;**12**:433-442

[6] Mathew GM, Puthiyamadam A, Sasikumar K, Ashoor S, Sukumaran RK. Biological treatment of prawn shell wastes for valorization and waste management. Bioresource Technology Reports. 2021;**15**:100788

[7] Pakizeh M, Moradi A, Ghassemi T. Chemical extraction and modification of chitin and chitosan from shrimp shells. European Polymer Journal. 2021;**159**: 110709

[8] Seenuvasan M, Malar CG, Growther L. Production of a biopolymer film from biological wastes and its statistical analysis. Bioresource Technology Reports. 2021;**13**:100610

[9] Sandford PA. Book review: Advances in Chitin Science, Volume VI. Carbohydrate Polymers. 2004;**56**:95-96

[10] Aiba S. Studies on chitosan: 2. Solution stability and reactivity of partially N-acetylated chitosan derivatives in aqueous media. International Journal of Biology. 1989;**11**: 249-252

[11] Knorr D. Use of chitinous polymers in food—A challenge for food research and development. Food Technology. 1984;**38**:85-97

[12] Hackman R. H. Studies on chitin: The action of mineral acids on chitin. Canberra: Division of Entomology C; 1962. pp. 526-537

[13] Winterowd JG, Sandford P. Chitin and chitosan. In: Alistair MS, editor. Food Polysaccharides and Their Applications. New York: Marcel Dekker; 1995. pp. 441-462

[14] Paterson M, Kennedy J. Book review: Chitin and chitosan. Sources, chemistry, biochemistry, physical properties and applications. Carbohydrate Polymers. 1990;**13**:116-117

[15] Joseph J, Manigundan K, Shamya Arokia Rajan M, Radhakrishnan M, Gopikrishnan V, Kumaran S, et al. Conversion of aquaculture waste into biomedical wealth: Chitin and chitosan journey. Advances in Materials Science and Engineering. 2022;**2022**:1-12

[16] Cho Y-W, Cho Y-N, Chung S-H, Yoo G, Ko S-W. Water-soluble chitin as a wound healing accelerator. Biomaterials. 1999;**20**(22):2139-2145

[17] Kritchenkov AS, Kletskov AV, Egorov AR, Tskhovrebov AG,

Kurliuk AV, Zhaliazniak NV, et al. New water-soluble chitin derivative with high antibacterial properties for potential application in active food coatings. Food Chemistry. 2021;**343**:128696

[18] Barikani M, Oliaei E, Seddiqi H, Honarkar H. Preparation and application of chitin and its derivatives: A review. Iranian Polymer Journal. 2014;**23**(4): 307-326

[19] Kandra P, Challa MM. Efficient use of shrimp waste: Present and future trends. Applied Microbiology and Biotechnology. 2011;**93**(1):17-29

[20] Kurita K. Chitin and chitosan: Functional biopolymers from marine crustaceans. Marine Biotechnology [NY]. 2006;**8**(3):203-226

[21] Abdou ES, Nagy KSA, Elsabee MZ. Extraction and characterization of chitin and chitosan from local sources. Bioresource Technology. 2008;**99**(5): 1359-1367

[22] Bigi F, Haghighi H, Siesler HW, Licciardello F, Pulvirenti A. Characterization of chitosan-hydroxypropyl methylcellulose blend films enriched with nettle or sage leaf extract for active food packaging applications. Food Hydrocolloids. 2021;**120**:106979

[23] Maleki G, Milani M. Functional Properties of Chitin and Chitosan-Based Polymer Materials: Handbook of Chitin and Chitosan. Netherlands, United Kingdom, United States: Elsevier Inc; 2020. pp. 177-198

[24] Roy JC. Solubility of Chitin: Solvents, Solution Behaviors and Their Related Mechanisms: s.l. IntechOpen; 2017. E-book ISBN: 978-953-51-4593-6

[25] Zhu K, Shi S, Cao Y, Lu A, Hu J, Zhang L. Robust chitin films with good biocompatibility and breathable properties. Carbohydrate Polymers. 2019;**212**:361-367

[26] Tsai GJ, Su W. Antimicrobial activity of shrimp chitin and chitosan from different treatments and applications of fish preservation. Fisheries Science. 2002;**68**:170-177

[27] Huang C, Chen S, Pan J. Optimal condition for modification of chitosan: A biopolymer for coagulation of colloidal particles. Water Research. 2000;**34**: 1057-1062

[28] Shahidi F, Abuzaytoun R. Chitin, chitosan and co-products: Chemistry, production, application and health effects. Advances in Food and Nutrition Research. 2005;**49**:93-135

[29] Minke R, Blackwell J. The structure of alpha-chitin. Journal of Molecular Biology. 1978;**120**(2):167-181

[30] Muzzarelli RAA. Natural Chelating Polymers. New York: Pergamon Press; 1973. p. 134

[31] Sannan T, Kurita K, Iwakura Y. Solubility change by alkaline treatment and film casting. Makromolekulare Chemie. 1975;**176**:1191-1195

[32] Sannan T, Kurita K, Iwakura Y. Studies on chitin. 2. Effect of deacetylation on solubility. Makromolekulare Chemie. 1976;**177**: 3589-3600

[33] Guo XF, Kikuchi K, Matahira Y, Sakai K, Ogawa K. Water-soluble chitin of low degree of deacetylation. Journal of Carbohydrate Chemistry. 2002;**21**(1-2): 149-161

[34] Aiba S, Izume M, Minoura N, Fujiwara Y. Studies on chitin. 2. Preparation and properties of chitin

membranes. Carbohydrate Polymers. 1985;**5**:285-295

[35] Muzzarelli RAA. Carboxymethylated chitins and chitosans. Carbohydrate Polymers. 1988;**8**:1-21

[36] Muzzarelli RAA, Muzzarelli C, Cosani A, Terbojevich M. 6-Oxychitins, novel hyaluronan-like regiospecifically carboxylated chitins. Carbohydrate Polymers. 1999;**39**:361-367

[37] Tokura S, Nishi N, Tsutsumi A, Somorin O. Studies on chitin. VIII. Some properties of water soluble chitin derivatives. Polymer Journal. 1983;**15**(6): 485-489

[38] Sun L, Du Y, Yang J, Shi X, Li J, Wang X, et al. F. Conversion of crystal structure of the chitin to facilitate preparation of a 6-carboxychitin with moisture absorption–retention abilities. Carbohydrate Polymers. 2006;**66**: 168-175

[39] Chen LY, Du Y. Aggregation behavior of *3,6-O*-carboxymethylated chitin in aqueous solutions. Journal of Applied Polymer Science. 2002;**86**: 1838-1843

[40] Kurita K, Inoue S, Nishimura S. Preparation of soluble chitin derivatives as reactive precursors for controlled modifications: Tosyl- and iodo-chitins. Journal of Polymer Science. 1991;**29**: 937-939

[41] Kim SJ, Kim S. Synthesis and characterization of ether-type chitin derivatives. Macromolecular Chemistry and Physics. 1994;**195**:1687-1693

[42] Park IK, Park YH. Preparation and structural characterization of water-soluble o-hydroxypropyl chitin derivatives. Journal of Applied Polymer Science. 2001;**80**:2624-2632

[43] Liu Y, Chen G, Hu KA. Synthesis, characterization and structural analysis of polylactide grafted onto water-soluble hydroxypropyl chitin as backbone. Journal of Materials Science Letters. 2003;**22**:1303-1305

[44] Nishimura T, Eto E, Yamada T. InventorProduction of Water-Soluble Chitin Oligomers. 1987

[45] Sashiwa H, Shigemasa Y. Chemical modification of chitin and chitosan 2: Preparation and water soluble property of N-acylated or N-alkylated partially deacetylated chitins. Carbohydrate Polymers. 1999;**39**:127-138

[46] Shigemasa Y, Usui H, Morimoto M, Saimoto H, Okamoto Y, Minami S, et al. Chemical modification of chitin and chitosan 1: Preparation of partially deacetylated chitin derivatives via a ring-opening reaction with cyclic acid anhydrides in lithium chloride/N,N-dimethylacetamide. Carbohydrate Polymers. 1999;**39**:237-243

[47] Vikhoreva GA, Gorbacheva IN, Gal'braikh LS. Synthesis and properties of water-soluble derivatives of chitin. A review. Fibre Chemistry. 1999;**31**: 274-278

[48] Kurita K, Kojima T, Nishiyama Y, Shimojoh M. Synthesis and some properties of nonnatural amino polysaccharides: Branched chitin and chitosan. Macromolecules. 2000;**33**: 4711-4716

[49] Vasnev VA, Tarasov A, Markova GD, Vinogradova SV, Garkusha OG. Synthesis and properties of acylated chitin and chitosan derivatives. Carbohydrate Polymers. 2006;**64**:184-189

[50] Nishi N, Nishimura S, Ebina A, Tsutsumi A, Tokura S. Preparation and

characterization of water-soluble chitin phosphate. International Journal of Biological Macromolecules. 1984;**6**: 53-54

[51] Philippova OE, Korchagina EV, Volkov EV, Smirnov VA, Khokhlov AR, Rinaudo M. Aggregation of some water-soluble derivatives of chitin in aqueous solutions: Role of the degree of acetylation and effect of hydrogen bond breaker. Carbohydrate Polymers. 2012; **87**(1):687-694

[52] Cho Y-W, Jang J, Park CR, Ko S-W. Preparation and solubility in acid and water of partially deacetylated chitins. Biomacromolecules. 2000;**1**(4):609-614

[53] Sugimoto M, Morimoto M, Sashiwa H, Saimoto H, Shigemasa Y. Preparation and characterization of water-soluble chitin and chitosan derivatives. Carbohydrate Polymers. 1998;**36**(1):49-59

[54] Kurita K, Koyama Y, Taniguchi A. Studies on chitin: IX. Crosslinking of water-soluble chitin and evaluation of the products as adsorbents for cupric ion. Journal of Applied Polymer Science. 1986;**31**(5):1169-1176

[55] Kurita K, Koyama Y, Nishimura S, Kamiya M. Facile preparation of water-soluble chitin from chitosan. Chemistry Letters. 1989;**9**:1597-1598

[56] Heux L, Brugnerotto J, Desbrieres J, Versali MF, Rinaudo M. Solid state NMR for determination of degree of acetylation of chitin and chitosan. Biomacromolecules. 2000;**1**(4):746-751

[57] Chi HL, Yeh AI, Pan MH, Chen SH. Physicochemical properties and film formation of the chitin hydrocolloid fabricated by a novel green process. Journal of Applied Polymer Science. 2021;**138**(31):50762

[58] Kurita K, Mori S, Nishiyama Y, Shimojoh M. N-alkylation of chitin and some characteristics of the novel derivatives. Polymer Bulletin. 2002;**48**: 159-166

[59] Claesson PM, Ninham BW. pH-dependant interaction between adsorbed chitosan layers. Langmuir. 1992;**8**: 1406-1412

[60] Anthonsen MW, Varum K. Solution properties of chitosans: Conformation and chain stiffness of chitosans with different degree of N-acetylation. Carbohydrate Polymers. 1993;**22**:193-201

[61] Piron E, Accominotti M, Domard A. Role of physical and physicochemical parameters on the kinetics of sorption. Langmuir. 1997;**13**(6):1653-1658

[62] Holme HK, Foros H, Pettersen H, Dornish M, Smidsrod O. Thermal depolymerization of chitosan chloride. Carbohydrate Polymers. 2001;**46**: 287-294

[63] Seoudi R, Nada AMA. Molecular structure and dielectric properties studies of chitin and its treated by acid, base and hypochlorite. Carbohydrate Polymers. 2007;**68**:728-733

[64] Kittur FS, Prashanth H, Sankar KU, Tharanathan RN. Characterization of chitin, chitosan and their carboxymethyl derivatives by differential scanning calorimetry. Carbohydrate Polymers. 2002;**49**:185-193

[65] Shen X, Ji Y, Yang Q, Zheng X. Preparation, characterization, and rheological properties of dibutyrylchitin. Journal of Macromolecular Science—Physics. 2010;**49**(2):250-258

[66] Sharma SK, Mulvaney SJ, Rizvi SSH. Food Process Engineering. Canada: John Wiley and Sons; 2000

[67] Chen B, Sun K, Zhang K. Rheological properties of chitin/lithium chloride, N, N-dimethyl acetamide solutions. Carbohydrate Polymers. 2004;**58**:65-69

[68] Taghizadeh SM, Davari G. Preparation, characterization and swelling behavior of N-acetylated and deacetylated chitosan. Carbohydrate Polymers. 2006;**64**(1):9-15

[69] Oh H, Kim Y. Antimicrobial characteristics of chitosan against food spoilage microorganisms in liquid media and mayonnaise. Bioscience, Biotechnology, and Biochemistry. 2001; **65**:2378-2383

[70] Hayes ER, Davies DH. Proceeding of First International Conference on Chitin and Chitosan. USA; 1978

[71] Noel TR, Ring SG, Whittam M. Glass transitions in low-moisture foods. Trends in Food Science and Technology. 1990;**1990**:62

[72] Zeleznak KJ, Hoseney RC. The glass transtion in starch. Cereal Chemistry. 1987;**64**:121-124

[73] Flink JM. In: Baglery EB, editor. Physical Properties of Foods. Connecticut: AVI; 1983. pp. 473-521

[74] Bhandari BR, Howes T. Implications of glass transition for the drying and stability of dried foods. Journal of Food Engineering. 1999;**40**:71-79

[75] Slade L, Levine H, Ievolella J, Wang M. The glassy state phenomenon in applications for the food industry: Application of the food polymer science approach to structure—Function relationships of sucrose in cookie and cracker systems. Journal of the Science of Food and Agriculture. 1993; **63**:133-176

[76] Slade L, Levine H. Water and the glass transition—Dependence of the glass transition on composition and chemical structure: special implications for flour funtionality in cookie baking. Journal of Food Engineering. 1994;**22**: 143-188

[77] Guinesi LS, Cavalheiro E. The use of DSC curves to determine the acetylation degree of chitin/chitosan samples. Thermochimica Acta. 2006;**444**:128-133

[78] Nieto JM, Peniche-Covas C. Characterization of chitosan by pyrolysis—Mass spectrometry, thermal analysis and differential scanning calorimetry. Thermochimica Acta. 1991; **176**:63-68

[79] Yilmaz E, Bengisu M. Preparation and characterization of physical gels and beads from chitin solutions. Carbohydrate Polymers. 2003;**54**:479-488

[80] Peeson M, Rujiravanit R, Supaphol P. Characterisation of beta-chitin/poly[vinyl alcohol] blend films. Polymer Testing. 2003;**22**:381-387

[81] Chassary P, Vincent T, Guibal E. Metal anion sorption on chitosan and derivative materials: A strategy for polymer modification and optimum use. Reactive and Functional Polymers. 2004; **60**:137-149

[82] Khalid MN, Agnely F, Yagoubi N, Grossiord JL, Couarraze G. Water state characterization, swelling behavior, thermal and mechanical properties of chitosan based networks. European Journal of Pharmaceutical Sciences. 2002;**15**(5):425-432

[83] Vogt BD, Soles CL, Lee HJ, Lin EK, Wu WL. Moisture absorption into ultrathin hydrophilic polymer films on different substrate surfaces. Polymer. 2005;**46**:1635-1642

[84] Slade L, Levine H. Non-equilibrium behavior of small carbohydrate water systems. Pure and Applied Chemistry. 1988;**60**:1841-1864

[85] Chang DS, Cho H. A development of food preservation with the waste of crab processing. Bulletin of the Korean Fisheries Society. 1989;**22**:70-78

[86] Darmadji P, Izumimoto M. Effect of chitosan in meat preservation. Meat Science. 1994;**38**(2):243-254

[87] Wang GH. Inhibition and Inactivation of Five Species of Foodborne Pathogens by Chitosan. Journal of Food Protection. 1992;**55**(11): 916-919

[88] Roberts GAF. Chitin Chemistry. London: The MacMilln Press Ltd; 1992

[89] A.O.A.C. Official Methods of Analysis. Washington DC: Chemist AoOA; 1990

[90] Singh M, Kanungo BK, Bansal TK. Kinetic studies on curing of hydroxy-terminated polybutadiene prepolymer-based polyurethane networks. Journal of Applied Polymer Science. 2002;**85**: 842-846

[91] Hayes ER, Davies DH. Characterization of chitosan. II: The determination of the degree of acetylation of chitosan and chitin. In: First International Conference on Chitin/Chitosan. Cambridge: Massachusetts Institute of Technology; 1978

[92] Winston PW, Bates DH. Saturated solutions for the control of humidity in biological research. JSTOR: Ecology. 1960;**41**:232-237

[93] Zerdin K. Antibacterial Study of Food Packaging Films. Sydney, Australia: CSIRO; 2007

[94] Kurita K, Sannan T, Iwakura Y. Evidence for formation of block and random copolymers of N-Acetyl-D-glucosamine and D-glucosamine by hetero- and homo-geneous hydrolyses. Makromolekulare Chemie. 1977;**176**: 3197-3202

[95] Kurita K, Ichikawa H, Ishizeki S, Fujisaki H, Iwakura Y. Studies on chitin: Modification reaction of chitin in highly swollen state with aromatic cyclic carboxylic acid anhydrides. Makromolekulare Chemie. 1982;**183**: 1161-1169

[96] Sreenivasan K. Thermal stability studies of some chitosan-metal ion complexes using differential scanning calorimetry. Polymer Degradation and Stability. 1996;**52**:85-87

[97] Tirkistani FAA. Thermal analysis of chitosan modified by cyclic oxygenated compounds. Polymer Degradation and Stability. 1998;**61**:161-164

[98] Tirkistani FAA. Thermal analysis of some chitosan Schiff bases. Polymer Degradation and Stability. 1998;**60**:67-70

[99] Despond S, Espuche E, Domard A. Water sorption and permeation in chitosan films: Relation between gas permeability and relative humidity. Journal of Polymer Science. 2001;**39**: 3114-3127

Chapter 3

Preparation and Bioactivity Applications of Novel Chitosan Derivatives

Mohsin Mohammed and Nadia Haj

Abstract

Chitosan (CS) is a substance abundant in nature. It is a biopolymer consisting of repetitive components of glucose and *N*-acetyl-glucose amine connected by (1,4)-glycosidic bonds. It has so many applications that are biodegradable, non-toxic, and biocompatible. The CS was loaded with 5-fluorouracil (5FU) *via* amide-mediated binding, and the resulting CSFUAC product was evaluated as a potential 5FU delivery agent. A new CS-Schiff base derivative was created using CS extracted from local fish scales by combining CS with another aromatic aldehyde. The antimicrobial effectiveness of the new product was evaluated. It includes two fungi and four strains of pathogenic bacteria. The MTT assay is employed to determine the cytotoxicity of the newly synthesized compounds. Finally, CS was used to synthesize a prodrug for colon cancer. As a colon cancer prodrug, methotrexate (MTX) was converted to the combined (methotrexate-imidazole) and linked with the CS to produce the CSMTX conjugate. Additionally, the compound's hemolytic action and chemical stabilities were evaluated. In the MTT, three types of cancer cell lines (MDAMB231, MCF7, and MDAMB453) were utilized to test how toxic the compounds made in the lab were to cancer cells.

Keywords: chitosan, chitosan-Schiff base, prodrug, cytotoxicity, antibacterial activity

1. Introduction

The chemical structure of chitin (SH)-derived chitosan (CS) has the chemical structure of "(1→ 4) 2-amino-2-deoxy—D-glucopyranose." This is a typical co-biopolymer origin in the shells of cockroaches, the shells of crustaceans, and fungal cellular walls (**Figure 1**). The primary sources of CS and SH are crustaceans-like crabs, shrimp, and fish scales. CS is the most abundant and superior natural substance in nature, and second only to cellulose. Due to its excellent quality and adaptability, CS is in a league. In addition, they possess unique properties, for example, non-toxic, mucosal adhesion, biodegradability, biocompatibility, antimicrobial activities, and hydrophilicity. However, it is cholesterol lowering. These characteristics make CS useful in medicine, horticulture, stabilizers for staple foods, biocatalysts, and biology [2–7].

The beneficial biological effects of CS include antitumor, antibacterial, and hemostatic effects and wound healing. Applications include biomedicine design,

IntechOpen

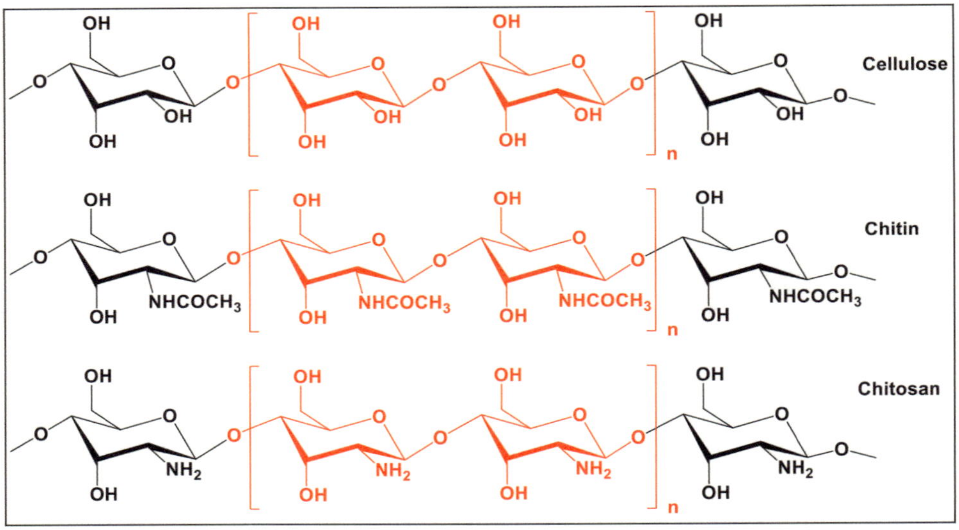

Figure 1.
Chitin, chitosan, and cellulose's molecular structures [1].

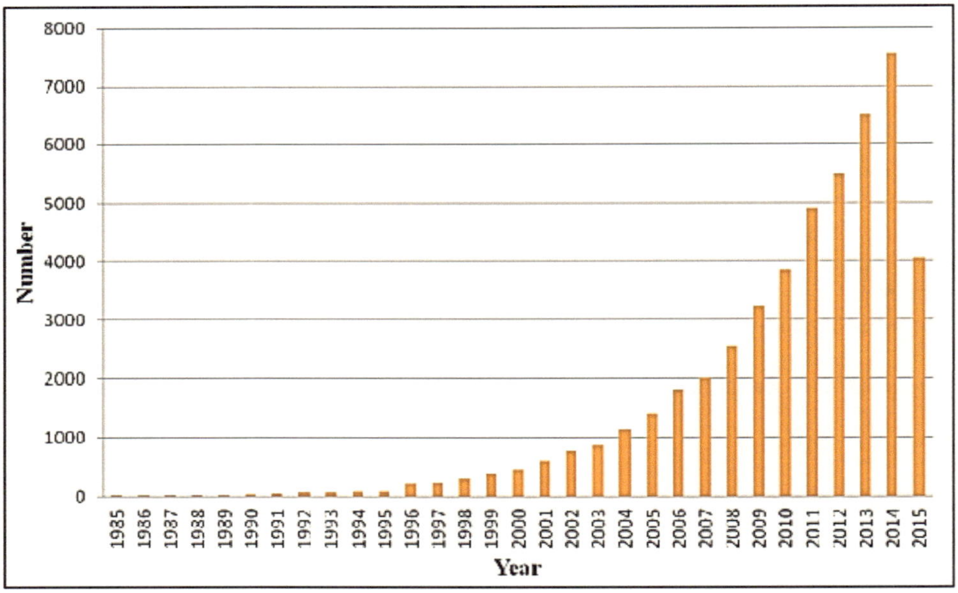

Figure 2.
Publications indexed by Scopus concerning chitosan and its derivatives.

pharmaceuticals, drug delivery, restoration materials, chelation of metal particles, water absorption processing, and plant safety [8–10]. From 1985 to June 2015, there was a significant increase in the research on chitosan and its derivatives; **Figure 2** depicts the number of Scopus-indexed publications on chitosan and its derivatives [11].

Figure 3 shows some chemical modifications of chitosan. **Table 1** illustrates some applications of chitosan derivatives.

Below is a summary of some applications of CS:

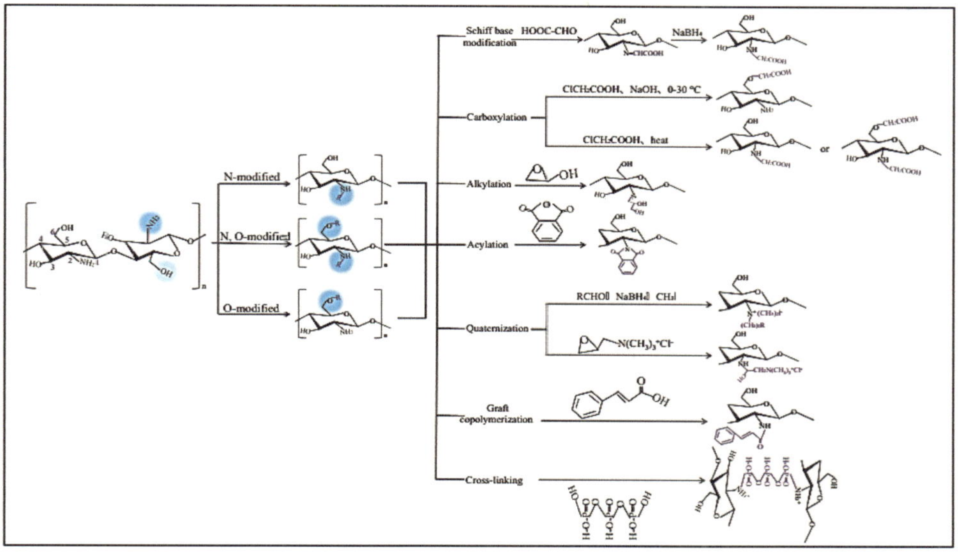

Figure 3.
Some chemical modifications of chitosan [12].

Fields	Applications	Descriptions/references
Food industry	Food packaging preservation	Film-forming property and antibacterial property [13].
	Food preservative	Antibacterial properties [14].
	Beverage clarifier	Flocculation [15].
	Wastewater treatment	Flocculation and chelating and adsorbing property [16].
Chemical industry	Chemical industry	Moisture absorption and water retention properties [17].
	Mouthwash	Antibacterial activity [18].
	CS fiber	Antibacterial and anti-wrinkle [19].
Textile industry	Dyeing and fixing	Physical adsorption and film-forming property [20].
	Coatings	Film-forming property and antibacterial property [21].
	Drug carrier	Film-forming, antioxidant, antitumor, and biocompatibility [22].
Medicine	Wound dressings	Anti-inflammatory and antibacterial properties [23].
	Tissue engineering	Proliferative, hemostatic, and antibacterial properties [24].
	Artificially simulate enzyme	Catalysis [25].
Functional materials	Liquid crystal materials	Optical and film-forming property [26].
Agriculture	Protect seed	Film-forming property and antibacterial property [27].
	Improve soil	Adsorption and bacteriostatic activity [28].
	Improve crops	Immune and bacteriostatic properties [29].

Table 1.
Examples of chitosan derivative applications.

2. Chitosan as a prodrugs for cancer

Worldwide, the leading cause of death is cancer. In 2008, cancer was the cause of death for about 13% of all people who died, or 7.6 million people. Numerous kinds of cancer are known, such as "prostate, colon, lung, and breast cancers," but colon cancer is the most lethal [30]. Numerous medications, such as Bevacizumab, (Avastin)Oxaliplatin, Folinic acid, 5-fluorouracil (5-FU), Methotrexate, and Celecoxib, are used to treat cancer and are used to treat colorectal cancer (Celecoxib). However, most drugs are consumed with food, restricting the higher gastrointestinal tract (GIT). They are not stable inside the body, which prevents the drug from concentrating effectively at the desired tumor site. Due to their lack of specificity, some medicines can also have the opposite effect [31]. 5-FU is a specific anticancer drug that is frequently employed in treatment. 5-FU has disadvantages, including rapid ingestion in the body, a small half-life, a propensity for activating cancer cell resistance and cytotoxicity. These facets necessitate higher medication dosages, which raises the danger of adverse properties. So, a perfect system for delivering 5-FU would send the drug in tiny amounts and let it out quickly at the target site [32–34]. A prodrug for colon cancer must meet several requirements, such as not being toxic, compatible with the body, and stable in the GIT [35–38]. This section designates the synthesis of a prodrug for the colon by binding CS with 5-FU. The drug was transformed to "5-fluorouracil-1-acetic acid (FUAC)," which was then conjugated to CS to produce "chitosan-1-acetic acid-5-fluorouracil (CS-FUAC)," which has been validated for colon cancer treatment. The CS was extracted from fish scales using a chemical process. Infrared Fourier transform and ultraviolet spectrophotometry was used to characterize the product. It was looked at as a possible delivery agent after a covalent bond was made between CS-FUAC and 5-FU.

2.1 SH and CS extraction

In Kirkuk, Iraq, fish scales were collected at a local fish market. Based on a documented procedure, for demineralization and deproteinization, a 1% sodium hydroxide and hydrogen chloride solution were prepared with the receptivity of 40 g/mol and 36.5 g/mol [39].

2.2 Preparation of CS-FUAC

As published in the literature, FUAC was constructed with minor variations. 5-FU and aq. of KOH are reacted and heated at 100°C for 90 minutes. Then, for over 6 hours, chloroacetic acid was added regularly using a water bath at 60°C while stirring. The result was acetified to produce needle crystals of FUAC with a yield of 60% [40]. To achieve "1-acetic acid-5-fluorouracilimidazoline," a solution of FUAC in DMSO and "1,1-carbonyldiimidazole" was combined with CS in a glacial acetic acid (GAA) aqueous solution (**Figure 4**). "FT-IR spectroscopy" was utilized to examine the CS conformation, and synthesized FUAC and the CS-FUAC. Using a UV spectrophotometer, the produced chemicals were analyzed. Experimentation was conducted at a wavelength of 273 nm. Using a "Bruker Avance (500) spectrometer", ^{1}H- and ^{13}C-NMR spectra were collected from DMSO-d6 and 1% CF_3COOD/D_2O solutions (**Figure 5**).

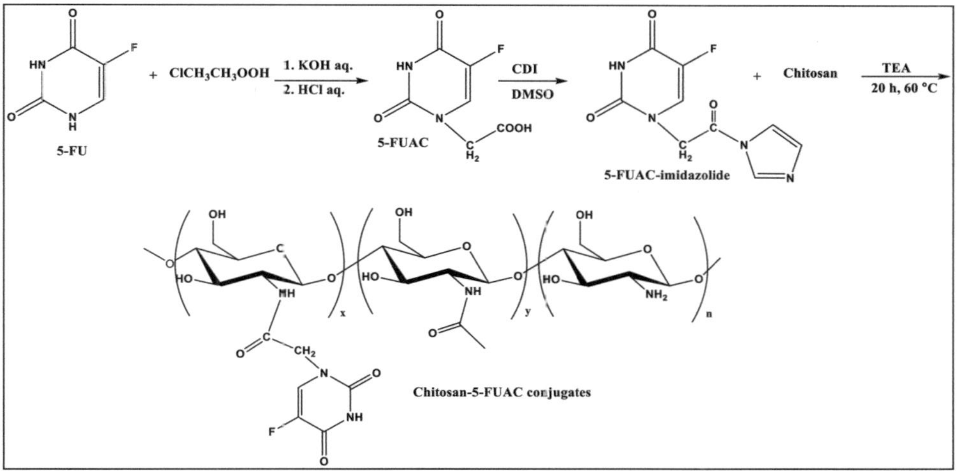

Figure 4.
Method for preparing "CS-FUAC."

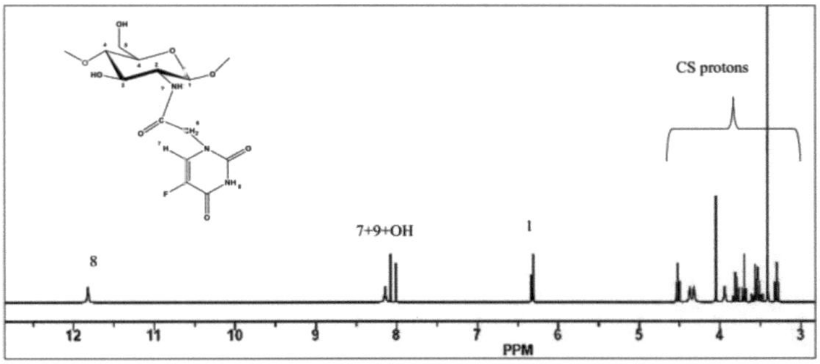

Figure 5.
^{1}H-NMR spectra of CS-FUAC.

2.3 The drug content percentage of CS-FUAC conjugates

UV-visible spectroscopy established the fraction of FUAC conjugated to CS. The amide link in the CS-FUAC conjugates was initially hydrolyzed in the primary medium. As predicted, the concentration of FUAC was raised by raising the molar fraction of CS to FUAC. **Table 2** displays the FUAC concentration in conjugates where the molar ratios of "CS: FUAC were 1:1 and 1:2," respectively.

2.4 The stability test of conjugates of CS-FUAC

During transit through GIT, the drug used in the treatment of colon cancer must be stable in various pH ranges. Therefore, two buffers with pH of 1.2 and 7.4 were used to examine the prodrug stability in acidic and basic environments to determine its performance in acidic and basic environments. Based on the way conjugates are released in an acidic buffer, only 3–5% could be released. In contrast, it was 3–4% in

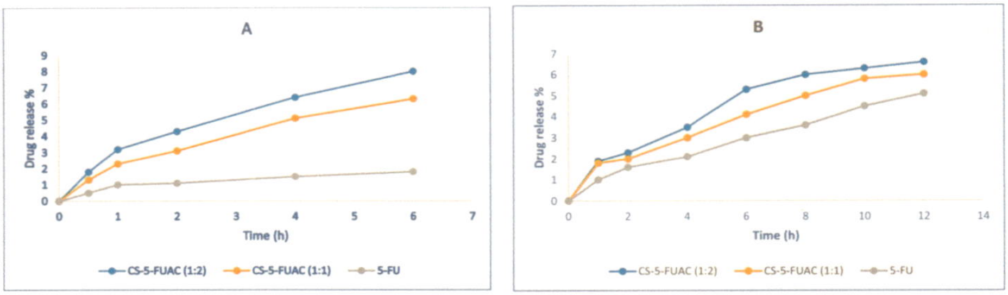

Figure 6.
Stability evaluation of CS-FUAC and 5-FU at pH 1.2 (A) and a pH of 7.4 (B).

Sample[i]	Drug loading (wt%)[ii]	Yield (%)
CS-FUAC (1:1)	0.16	60
CS-FUAC (1:2)	0.75	68

[i]*All reaction was carried out at 75°C for 24 h.*
[ii]*Drug loading determined by UV.*

Table 2.
Reaction data for CS-FUAC.

the primary buffer (**Figure 6**). In acidic environments, drug release is more significant than in neutral conditions. This could be because of the hydrolysis of an amide linkage in the acidic condition.

2.5 Conclusion

CS-FUAC, a potential colon-prodrug, was synthesized. The created conjugates appeared more stable under the initial conditions. In addition, CS-FUAC was more durable than the original medication in different pH conditions. An *in vitro* cytotoxicity investigation revealed that those synthesized derivatives are more active than free drugs. The cytotoxicity test, *in vitro*, demonstrates that these prodrugs are significantly more effective against "human colorectal cancer cell lines (HT-29)" than free medicine. In addition, it was nearly twice as cytotoxic to colon cancer cells compared with normal cells. Considering these outcomes, CS-FUAC as a prodrug appears to be an excellent method for colon delivery.

3. Synthesis and evaluation of Schiff base chitosan in biological systems

The Schiff base reactions produce the essential CS derivatives because of their application in the organic field. Furthermore, the response of CS with aromatic rings or heterocyclic aldehydes has led to the formation of stable Schiff bases (SBs), which are excellent molecules with uses in pharmacology, medicine, and other fields. Antimicrobial and cancer-prevention medications, for example [41, 42]. Interaction between CS amine sites and aldehydes or ketones, then removing water molecules, produces CSSBs [43]. Quinoline and quinazoline derivatives can also be found in various natural products, producing heterocyclic molecules with critical pharmacological applications. Quinoline and its results have antimalarial, antiviral, antibacterial, analgesic, anti-hepatoma, and anti-inflammatory properties. Oxazole derivatives are considered essential heterocyclic molecules in medicinal chemistry [44–56]. CH is extracted from a fish scale described previously and deacetylated to produce CS. This study's objective is to estimate the DD percentage of CS, which is carried out by acid-base titration for samples of CS collected at various stages of the manufacturing process. In this investigation, FTIR and TGA will be used to characterize and validate the physicochemical parameters of the products. This investigation seeks to develop a synthetic method for three new CS Schiff bases (CSSB)s compounds by combining CS with "2-chloroquinoline-3-carbaldehyde, quinazoline-6-carbaldehyde, and oxazole-4-carbaldehyde." Furthermore, the antibacterial possible of CS and its novel derivatives were investigated using two types of "fungi, C. Albicans and A. fumigate." Utilizing FT-IR, (^{1}H, and ^{13}C) NMR spectroscopy, the structures of the manufactured products were established. The cytotoxicity of newly synthesized derivatives was determined using the MTT experiment.

3.1 Method of acid-base titration

The acid-base titration technique was used to calculate a CS sample's degree of deacylation (DD). An indicator was used to control the endpoint, which turned to a blue-green color. The endpoint of the acid-base titration was used to calculate the DD percentage. The below equations were used to calculate the DD% of the deacetylated SH [57].

$$(-\text{NH2\%}) = \left(\frac{0.016[(\text{C1V1}) - (\text{C2V2})]}{\text{Wx100}}\right. \quad (1)$$

$$(\text{DD\%}) = 203(\text{NH2\%}/(16 + 42(-\text{NH2\%}))\text{x100} \quad (2)$$

where the concentration and volume of HCl used are C1 and V1, the concentration and volume of NaOH used for titration are C2 and V2, and the weight of samples used for acid-base titration is W.

3.2 Synthesis of CSSBs

CSSBs were created by dissolving extracted CS in 2.0 percent aq. acetic acid, and carbonyl compounds dissolved in ethanol and added to the same amount of "CS. CS-P1, CSP2, and CS-P3" were the products of reactions between CS and "2-chloroquinoline-3-carbaldehyde, quinazoline-6-carbaldehyde, and oxazole-4-carbaldehyde", respectively. Three CSSBs were synthesized, as exposed in **Figure** 7.

The configuration of the synthesized derivatives was established with "FT-IR spectroscopy, ^{1}H NMR, and ^{13}C NMR." All FT-IR spectra of the newly prepared derivatives presented a band between 1633 and 1655 cm^{-1}, related to the (-C=N) bond. Bands between 1400 and 1500 cm^{-1} and 1057 cm^{-1} correspond to (C-C) and in-plane (C-H) bonds. The absence of a band in the region between 1660 and 1730 cm^{-1} showed that the carbonyl group was absent, denoting that no free carbonyl remained. The (C-H) stretching of (CH_3- and -CH_2-) is represented by the vibrational bands at 2921 and 2883 cm^{-1}. Bands at 1155 and 900 cm^{-1} were found in the glycosidic bonds. The glycosidic bonds and the (C-O, C-C, and C-O-C) stretching of the glycan ring

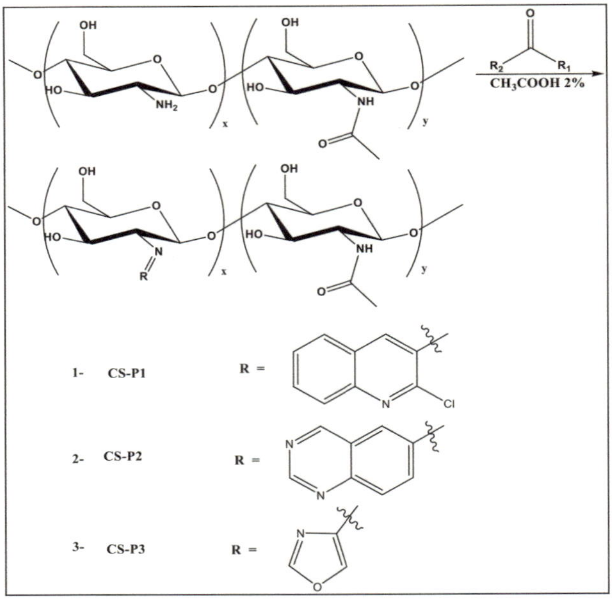

Figure 7.
Development of CSSB derivatives.

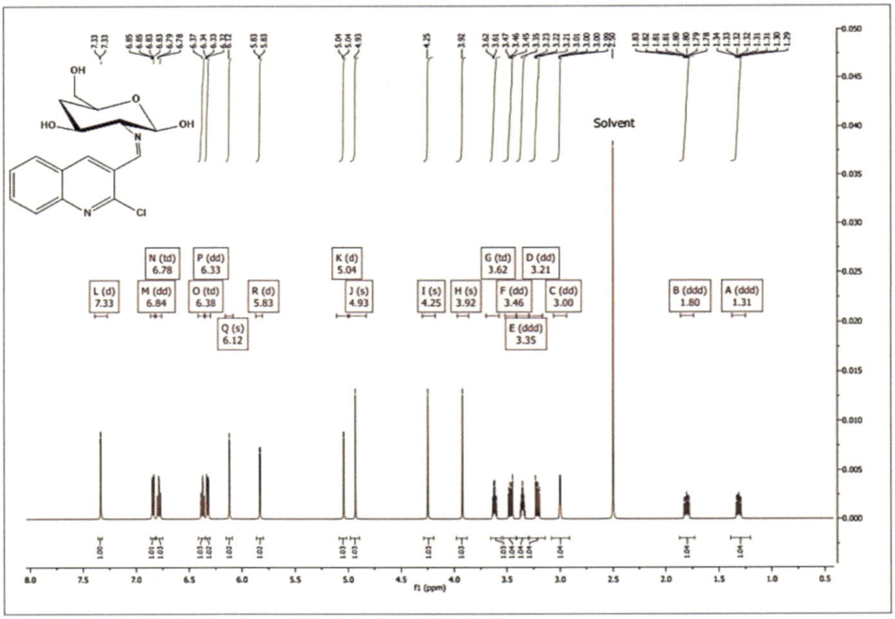

Figure 8.
¹H NMR form CS-P1 compound.

were linked to the 1205–975 cm^{-1} bands [58, 59]. "Using ^{1}H and ^{13}C NMR, the structure of the prepared CSSBs was confirmed." The ^{1}H NMR spectra of the synthesized derivatives CS-P1and CS-P2 are depicted in **Figures 8** and **9**, respectively. The ^{13}C NMR spectra for CSP1 is illustrated in **Figure 10**.

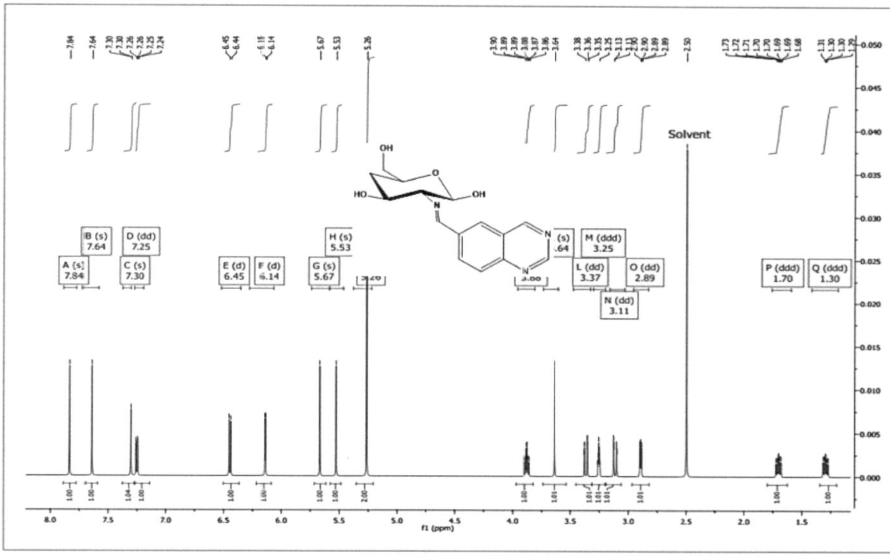

Figure 9.
^{1}H NMR spectra of CS-P2.

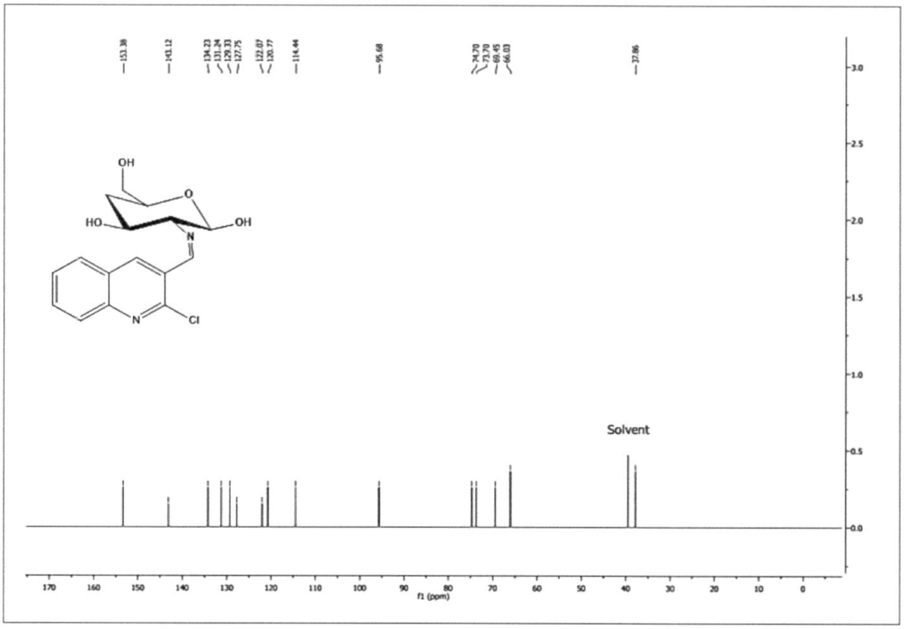

Figure 10.
^{13}C NMR spectra CS-P1.

3.3 Examine the solubility

The solubility of the synthesized products was investigated using a variety of organic solvents. DMSO and mixtures of DMSO-CF_3COOH in equal proportions were used. The results are shown in **Table 3**. Certain solvents, such as diluted HCl and

Solvents											
	CH_3COOH		CF_3COOH		DMSO	HCl	NaOH	H_2O	KOH	DMSO+ CF_3COOH
Comp. codes	25°C	70°C	25°C	70°C	25°C	25°C	25°C	25°C	25°C	25°C
CS-P1	S+	S*	S**	S**	S	S*	S+	S+	S**	S
CS-P2	S+	S*	S**	S**	S	S*	S+	S+	S**	S
CS-P3	S+	S*	S**	S**	S	S*	S+	S+	S**	S

S = soluble, S+ = insoluble, S* = partially soluble and swelling, S** = partially soluble.

Table 3.
CSSB solubility characteristics in a range of solvents.

CH_3COOH, showed incomplete dissolution or swelling at 70°C. Most inorganic solvents are insoluble in the products.

3.4 *In vitro* cytotoxicity examination

The MTT test is a colorimetric test used to measure cytotoxicity and cell viability. Based on MTT, the cytotoxicity of the synthesized derivatives was assessed, as depicted in **Table 4**.

Compared to the control, the outcome of the verified derivatives 'CS-P1, CS-P2, and CS-P3' exhibit the minimal difference between them. Several previous types of research have confirmed that CS and CSSB derivatives are non-toxic to cells. Consequently, CS has numerous medical applications [60–62].

3.5 Antimicrobial assessment

The CSSB derivatives' antibacterial activity was assessed using the inhibition zone technique. **Table 5** illustrates the results. The outcomes show that CS and all CSSBs affect *E. coli* and *K. pneumonia* strains as CS. According to the study, CS can prevent *S. aureus* and *E. coli* from forming new cells [63]. CSSBs derivatives presented antimicrobial activity against *S. aureus*, with inhibition zones of "22±0.3, 20±1.2, and 19±0.62 mm" for "CS-P1, CS-P2, and CS-P3," respectively. The synthesized compounds also have antibacterial activity against *S. mutans*, with inhibition zones of "15±0.89, 17±0.50, and 18±1.20 mm," respectively. Two fungal strains were used to test the CSSBs' antifungal activity, and all the CSSBs tested showed positive results.

Comp. conc. (mg)	Viable cells in the presence of CS	Viable cells in the presence of CS-P1	Viable cells in the presence of CS-P2	Viable cells in the presence of CS-P3
25	99 ± 0.83	99 ± 0.73	99 ± 0.60	99 ± 0.75
50	97 ± 0.73	99 ± 0.91	98 ± 1.3	99 ± 0.89
100	94 ± 0.63	98 ± 0.74	98 ± 1.2	98 ± 0.50
150	93 ± 0.88	97 ± 0.65	98 ± 0.62	96 ± 1.2
200	89 ± 0.53	90 ± 0.72	90 ± 0.74	91 ± 1.2

The experiment was repeated three times and mean was calculated.

Table 4.
The CS and SB derivative cytoxicity assessment.

	Gram-negative bacteria		Gram-positive bacteria		Fungi	
Comp. codes	***E. coli***	***K. pneumonia***	***S. aureus***	***S. mutans***	***C. albicans***	***Ammophilus fumigatus***
CS	24 ± 0.63	26 ± 0.73	NA	NA	26 ± 0.79	16 ± 0.83
CS-P1	22 ± 0.73	28 ± 0.91	22 ± 0.3	15 ± 0.89	34 ± 0.99	26 ± 0.91
CS-P2	27 ± 0.83	27 ± 0.72	20 ± 1.2	17 ± 0.5	31 ± 1.29	25 ± 0.72
CS-P3	22 ± 0.98	26 ± 0.65	19 ± 0.62	18 ± 1.2	26 ± 0.49	21 ± 0.65

NA means not detected.

Table 5.
Results of antibacterial and antifungal action of CS and CSSBs.

According to the published studies, numerous mechanisms are expected to explain how CS acts on bacteria, which vary depending on the metabolic process, the type of microorganism, and the cell wall composition. The initial recommendation is to disrupt the organism's cell wall electrostatic attraction between the positively charged amine in CS and the negatively charged residue group in bacterial nucleic acid cellular components such as COO^- or $PO4^{-2}$. The interface of bacterial DNA with CS is proposed as the second procedure. Protein and messenger RNA perversion into bacterial cells is induced by CS, followed by nuclei. Another theory relies on the ability of CS to form metal complexes. Metal complexes such as Zn^{2+}, Mg^{2+}, and Ca^{2+} are examples. Metals are required for bacterial metabolic and growth processes [63].

3.6 Conclusion

The hunt for new antibiotics has risen in tandem with the rise in antibiotic-resistant microorganisms. CS may be good material in this field. Cyprinus scales extracted 89% of the CS from local market-purchased Carpio fish. Acid-base titration was used to determine the DD % and describe how CS is put together. Three new CSSB derivatives that included the branches' distinct parts were synthesized, and their formations were confirmed using "FT-IR and ^1H and ^{13}C NMR spectroscopy." The new SB configuration was antibacterial against many bacteria and fungi tested. CSSBs had practically no effect on cytotoxic mouse fibroblast cell lines after being prepared. As a result of the findings above, it is reasonable to believe that the prepared CSSBs could be utilized with high efficiency, care, and performance in numerous biomedical fields.

4. A novel prodrug of methotrexate based on chitosan and evaluation of their bioactivity

Methotrexate (MTX) is the most effective medicine for treating several types of cancer, including colon cancer. Nevertheless, this medication can reduce the bioavailability of the goal material. It is administered orally and rapidly digested. MTX is an antimetabolic agent that inhibits folic acid metabolism. 1948 marked the beginning of clinical use of the drug as an anticancer agent, following the finding of its abnormal effects on DNA combination [64]. Due to the drug's physical and chemical characteristics, oral administration results in sedate retention at the beginning of the gastrointestinal tract. The drug's numerous significant limitations and disadvantages were demonstrated once aimed directly at a specific position of absorption or a particular portion of the alimentary channel, for instance, the colon. As a result, it is critical to explain how the medicine arrived at a specific treatment location [65]. Recent preparation of the prodrug involved covalently attaching the colon drug to the carrier. These prodrugs frequently alter their physical and chemical characteristics to improve infatuation at the activity site, increase the duration of action, and reduce toxicity and side effects [66]. This project is designed to develop a prodrug for colon cancer treatment by incorporating MTX into a biopolymer. The MTX was then converted to "methotrexate – imidazole" and loaded into CS to create colon cancer prodrugs containing CS-MTX conjugates. The structure of the synthesized derivatives was confirmed using spectroscopic analysis. The compound's chemical stability and hemolytic activity were also investigated. The drug concentration percentages were calculated. The "MDA-MB-231, MCF-7, and MDA-MB-453" cell lines were used to test the cytotoxicity of the prepared derivatives in a dish with MTT.

4.1 Synthesis of CS-MTX

The CS-MTX compound was developed as a possible treatment for MTX-induced colon cancer. As previously described, the CS was extracted and used to prepare the CS-MTX. As shown in **Figure 11**, methotrexate and imidazole were used to kickstart the process. To make MTX-imidazole *in situ*, equal quantities of "MTX and imidazole" were added to the mixture. A small quantity of "*N*, *N*-carbonyl diimidazole (CDI)" was added. At a concentration of 2%, CS was dissolved in GAA. The catalyst, triethylamine (TEA), was added in a few drops [67].

The confirmation of the CS-MTX compounds was established by "NMR and FTIR spectroscopy." A downfield signal at 6.02 Hz on the ^{1}HNMR chart indicated the proton was linked to C1 of the sugar (**Figure 12**). The proton signal was observed at 3.60, which is connected to C1. Since the two protons were in diverse environments, the methylene group C5 showed signals at 2.53, and C6 led signals at 2.08 and 2.28. The carbonyl signals were shown in the ^{13}C NMR spectrum (**Figure 13**) at 173.2 and 173.4. The value of the anomeric carbon was 101.5. The IR spectrum of the conjugate (**Figure 14**) revealed the CS and MTX vibrations. The N-H group had peaks at 3064 cm^{-1}, while the C-H group had 3032–2834 cm^{-1}. At 1660 and 1720 cm^{-1}, COOH group C-O vibration-related signals were detected. The (C-C, C-N, and CO) vibrations were found in signals between 1499 and 1183 cm^{-1}. A new peak at 1700 cm^{-1} confirmed the development of an amide bond between MTX and CS. This bond was vulnerable to acid hydrolysis or the amidase enzyme, allowing the drug to be released.

4.2 Determination of drug contains

The percentage of MTX-CS was calculated using UV-visible spectroscopy. The initial conditions for this experiment were based on the amide bond hydrolysis between MTX and CS. As expected, increasing the amounts of CS and MTX in the conjugates improved the results. The CS-MTX (1:1) contained 0.60% w MTX, and the yield was 65.0%.CS-MTX (1:2), on the other hand, had a content of 0.72% w and produced 70%. All reactions were carried out at 50°C for 20 hours. The proportion of medicine left was determined using this equation [68]:

$$\%\left(\frac{\mathrm{w}}{\mathrm{w}}\right)\text{of MTX loading} = \frac{\text{MTX amount}}{\text{CS} - \text{MTX conjugates amount}} \times 100 \quad (3)$$

4.3 Hemolytic exercise

The percentage of hemolysis produced by CS-MTX derivatives was investigated at various concentrations. The proportion of hemolysis increased as the concentration of CS-MTXs increased, as shown in **Figure 15**. According to the test results, the CSMTXs'

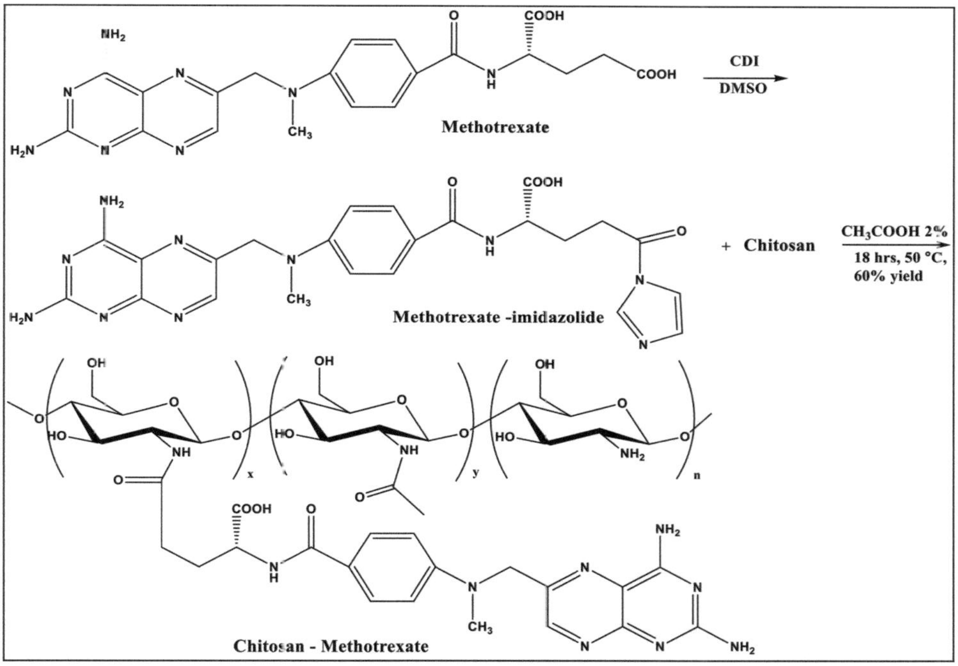

Figure 11.
Synthesis of CS-MTX.

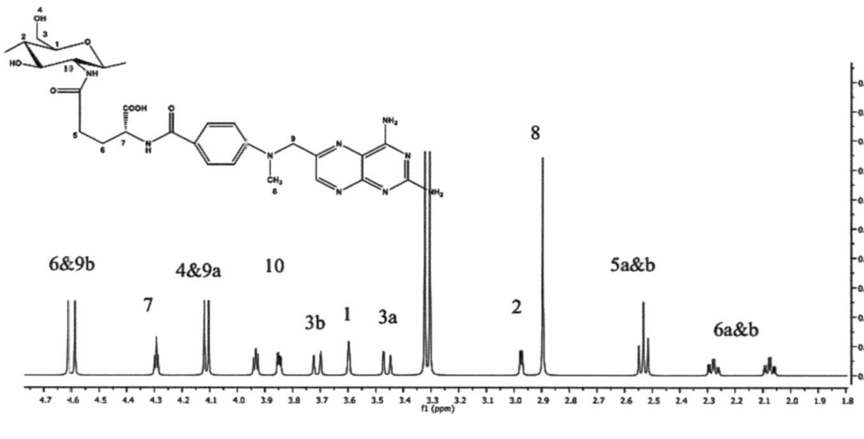

Figure 12.
The ^{1}H NMR of "CS-MTX."

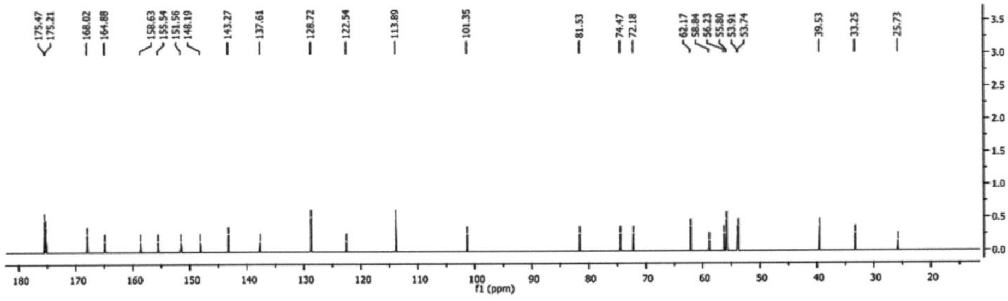

Figure 13.
The ^{13}C NMR of "CS-MTX."

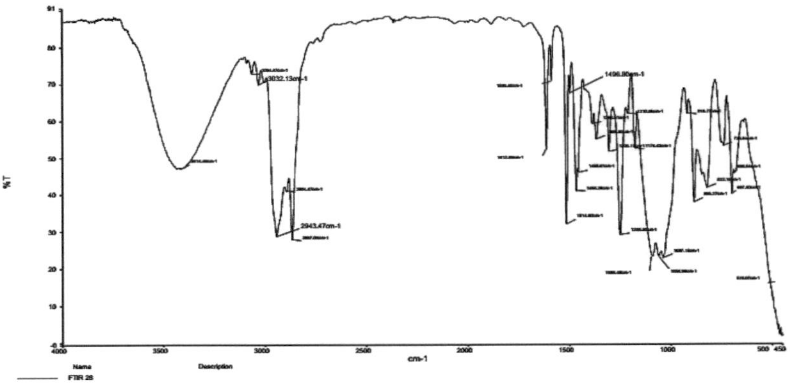

Figure 14.
The FT-IR of the "CS-MTX."

hemolysis rate was less than 4.5 percent, below the international standard of less than 5% [69]. This assessment was conducted depending on the method described [70]. The rate of hemolysis increases as the level of CS-MTX rises. The hemolysis rate in the CSMTXs was less than 4.5 percent, less than the international standard of less than 5 percent. White rabbit red blood cell (RBC) samples were used in this experiment. A 0.9 percent of NaCl solution was added after 3 mL of blood was extracted and centrifuged for 20 minutes at 4000 rpm. The controls were made by mixing 1 mL of red blood cells and distilled water with 5 mL of normal saline. A sequence of solutions containing both derivatives was carried out by adding 6,2,0.6,0.3,0.7 g of MTX and the prepared combination to tubes, followed by 2 mL of normal saline and 1 mL of red blood cells. The resulting solutions were kept for 2 hours at 37°C in a water bath. Separately, the tube was centrifuged at 4000 rpm for 20 minutes. At a wavelength of 541 nm, UV absorbance was measured. The following equation used to calculate the percentage of hemolysis was as follows:

$$\%\text{Hemolysis} = \frac{A\text{ sample} - A\text{ negative control}}{A\text{ positive control} - A\text{ negative control}} \times 100 \qquad (4)$$

4.4 Intracellular cytotoxicity

Three human breast cancer cell lines were used in this project "MDA-MB-231, MCF-7, and MDA-MB-453." The cell lines' viability was decreased by CSMTX

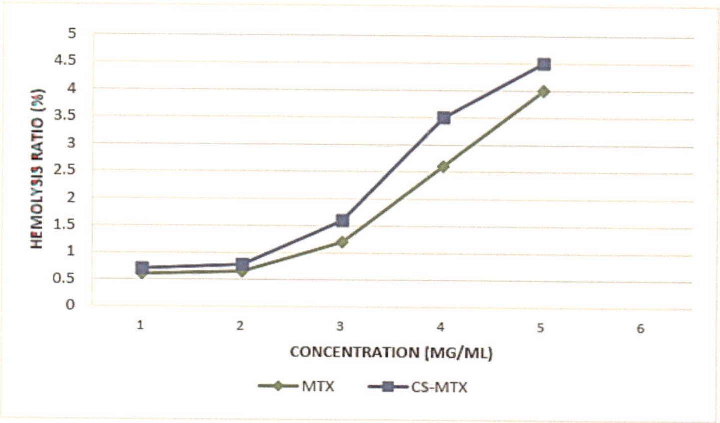

Figure 15.
The hemolysis caused by 'CS-MTX' treatment.

conjugates in a dose-dependent manner, according to MTT analyses. CS-MTX cytotoxicity was variable across all cell lines tested. The IC50 values for "MCF-7, MDA-MB-231, and MDA-MB-453 were 363.5±31.2, 198.8±20.4, and 163.4±10.8 g of CS-MTX/mL," respectively, over 24 hours. However, the cell "MDA-MB-453" used CS-MTX more precisely than the other cell lines in **Figure 16**. The cytotoxicity of CS MTX and the free medicine MTX was studied using "MCF-7, MDA-MB-231, and MDA-MB-453 cells," and the effect was calculated using the MTT assay [71]. A 96-well plate was used to test a range of CS-MTX and MTX concentrations (1–10 M). These concentrations were added before coating the plate with roughly 3 x 103 cells. After that, the plate was kept for 12, 24, and 48 hours. After removing the previous media, new

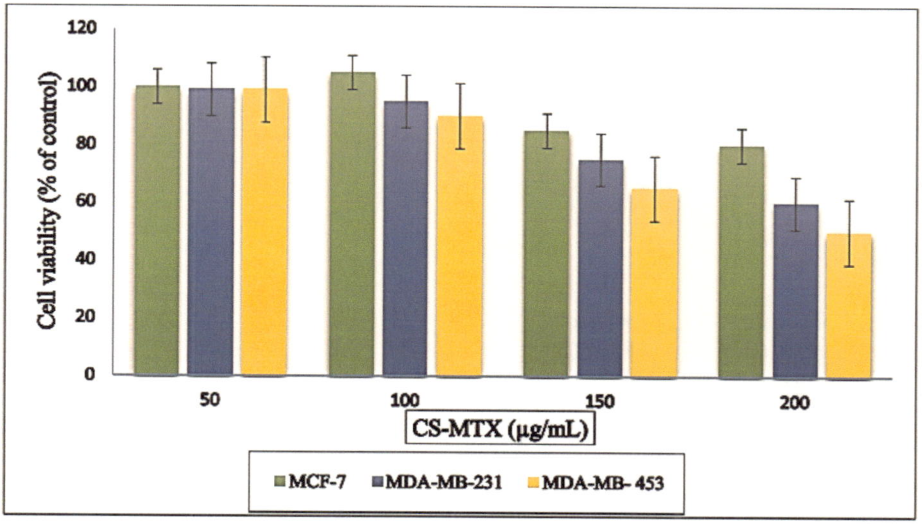

Figure 16.
MTT viability tests cell lines after mixing 200 g/mL in DMSO as a control for 24 hours. "Data were calculated as mean SD (n = 3) with a P-value of 0.01 compared to the control." Each solution's final exhibition was equal to 0.5 percent.

media were added, including (DMEM/F12) supplemented with 15 L of (MTT, concentration—500 g/mL). The cells were then grown at 37°C for 3–4 hours. The dark-blue formazan crystals were dissolved in DMSO. A Cytation three multimode plate reader manufactured in the United States was used to calculate their absorptivity at 574 nm for each well. The equation below was used to convert the obtained absorbance values into applied rate cells for unprocessed control cells:

$$\text{Relative cell viabillity} = \frac{\text{Absorbance of the sample}}{\text{Absorbance of the control}} \times 100 \quad (5)$$

4.5 Conclusion

The synthesized CS-MTX derivatives are most likely human colon cancer prodrugs. Under acidic conditions, the prepared derivatives were satisfactorily stable at a pH of 1.2. The synthesized compounds had a long half-life value of 4.52 in acidic conditions and 16.01 in basic media than the original drug. The CS-MTX conjugate produced significant results in the MTT assay. The three "cancer cell lines MCF-7, MDA-MB-231, and MDA-MB-453" showed a dose-dependent reduction in viability compared to the origin medicine. The IC50 values were "363.53±1.2, 198.82±0.4, and 163.41±0.8 g of CS-MTX/mL after 24 hours." These findings suggest that CS-MTX as a prodrug could be helpful in the treatment of colon cancer. Additional tests are being conducted to estimate the synthesized prodrug's various biological activities.

Author details

Mohsin Mohammed* and Nadia Haj
University of Kirkuk, Kirkuk, Iraq

*Address all correspondence to: althker1@uokirkuk.edu.iq

References

[1] Mohammed MO, Hussain KS, Haj NQ. Preparation and bioactivity assessment of chitosan-1-acetic acid-5-flurouracil conjugates as cancer prodrugs. Molecules. 2017;**22**:1629

[2] Haj NQ, Mohammed MO, Mohammood LE. Synthesis and biological evaluation of three new chitosan Schiff base derivatives. ACS Omega. 2020;**5**:13948-13954

[3] Rodríguez-Rodríguez R, Espinosa-Andrews H, Velasquillo-Martínez C, et al. Composite hydrogels based on gelatin, chitosan and polyvinyl alcohol to biomedical applications: A review. International Journal of Polymeric Materials and Polymeric Biomaterials. 2020;**69**:1-20

[4] Muhd Julkapli N, Akil HM, Ahmad Z. Preparation, properties and applications of chitosan-based biocomposites/blend materials: A review. Composite Interfaces. 2011;**18**:449-507

[5] Giannakas A, Patsaoura A, Barkoula N-M, Ladavos A. A novel solution blending method for using olive oil and corn oil as plasticizers in chitosan based organoclay nanocomposites. Carbohydrate Polymers. 2017;**157**: 550-557

[6] Casadidio C, Peregrina DV, Gigliobianco MR, et al. Chitin and Chitosans: Characteristics, eco-friendly processes, and applications in cosmetic science. Marine Drugs. 2019; **17**:369

[7] Elieh-Ali-Komi D, Hamblin MR. Chitin and chitosan: Production and application of versatile biomedical nanomaterials. International Journal of Advanced Research. 2016;**4**: 411-427

[8] Molnár Á. The use of chitosan-based metal catalysts in organic transformations. Coordination Chemistry Reviews. 2019;**388**:126-171

[9] Tang X, Huang T, Zhang S, et al. The role of sulfonated chitosan-based flocculant in the treatment of hematite wastewater containing heavy metals. Colloids Surfaces A Physicochemical and Engineering Aspects. 2020;**585**: 124070

[10] Matica A, Aachmann FL, Tøndervik A, Sletta H, Ostafe V. Chitosan as a wound dressing starting material: Antimicrobial properties and mode of action. International Journal of Molecular Sciences. 2019;**20**:5889

[11] Cheung RCF, Ng TB, Wong JH, et al. Chitosan: An update on potential biomedical and pharmaceutical applications. Marine Drugs. 2015;**13**: 5156-5186

[12] Wang W, Xue C, Mao X. Chitosan: Structural modification, biological activity and application. International Journal of Biological Macromolecules. 2020;**164**:4532-4546

[13] Uranga J, Puertas AI, Etxabide A, et al. Citric acid-incorporated fish gelatin/chitosan composite films. Food Hydrocolloids. 2019;**86**:95-103

[14] Khan I, Tango CN, Miskeen S, et al. Evaluation of nisin-loaded chitosan-monomethyl fumaric acid nanoparticles as a direct food additive. Carbohydrate Polymers. 2018;**184**:100-107

[15] Lia ZX, Zhoub XY, Tian ZH, et al. Application of modified chitosan in fruit juice clarification. Applied Mechanics and Materials. 2014;**651-653**:211-214

[16] Kavitha E, Sowmya A, Prabhakar S, et al. Removal and recovery of heavy metals through size enhanced ultrafiltration using chitosan derivatives and optimization with response surface modeling. International Journal of Biological Macromolecules. 2019;**132**: 278-288

[17] Chen K, Guo B, Luo J. Quaternized carboxymethyl chitosan/organic montmorillonite nanocomposite as a novel cosmetic ingredient against skin aging. Carbohydrate Polymers. 2017;**173**: 100-106

[18] Rangrazi A, Bagheri H, Ghazvini K, et al. Synthesis and antibacterial activity of colloidal selenium nanoparticles in chitosan solution: A new antibacterial agent. Materials Research Express. 2019; **6**:1250h3

[19] Judawisastra H, Hadyiswanto IOC, Winiati W. The effects of demineralization process on diameter, tensile properties and biodegradation of chitosan fiber. Procedia Chemistry. 2012; **4**:138-145

[20] Roy J, Salaün F, Giraud S, et al. Chitosan-based sustainable textile technology: Process, mechanism, innovation, and safety. In: Biological Activities and Application of Marine Polysaccharides. London: InTech; 2017. DOI: 10.5772/65259

[21] Costa EM, Silva S, Madureira AR, et al. A comprehensive study into the impact of a chitosan mouthwash upon oral microorganism's biofilm formation in vitro. Carbohydrate Polymers. 2014; **101**:1081-1086

[22] Song M, Li L, Zhang Y, et al. Carboxymethyl-β-cyclodextrin grafted chitosan nanoparticles as oral delivery carrier of protein drugs. Reactive and Functional Polymers. 2017;**117**:10-15

[23] Haitao Y, Yifan C, Mingchao S, et al. A novel polymeric nanohybrid antimicrobial engineered by antimicrobial peptide MccJ25 and chitosan nanoparticles exerts strong antibacterial and anti-inflammatory activities. Frontiers in Immunology. 2022;**12**:5955

[24] Cao S, Xu G, Li Q, et al. Double crosslinking chitosan sponge with antibacterial and hemostatic properties for accelerating wound repair. Composites. Part B, Engineering. 2022; **234**:109746

[25] Lyu X, Gonzalez R, Horton A, et al. Immobilization of enzymes by polymeric materials. Catalysts. 2021;**11**:1211

[26] Xie SM, Yuan LM. Recent development trends for chiral stationary phases based on chitosan derivatives, cyclofructan derivatives and chiral porous materials in high performance liquid chromatography. Journal of Separation Science. 2019;**42**:6-20

[27] Chandra S, Chakraborty N, Panda K, et al. Chitosan-induced immunity in *Camellia sinensis* (L.) O. Kuntze against blister blight disease is mediated by nitric-oxide. Plant Physiology and Biochemistry. 2017;**115**:298-307

[28] Chang H-q, Wang Q-z, Li Z-j, Wu J, Xu X-f, Shi Z-y. The effects of calcium combined with chitosan amendment on the bioavailability of exogenous Pb in calcareous soil. Journal of Integrative Agriculture. 2020;**19**:1375-1386

[29] Gabriel Paulraj M, Ignacimuthu S, Gandhi MR, et al. Comparative studies of tripolyphosphate and glutaraldehyde cross-linked chitosan-botanical pesticide nanoparticles and their agricultural applications. International Journal of Biological Macromolecules. 2017;**104**: 1813-1819

[30] Sharma R, Rawal RK, Gaba T, et al. Design, synthesis and *ex vivo* evaluation of colon-specific azo based prodrugs of anticancer agents. Bioorganic & Medicinal Chemistry Letters. 2013;**23**: 5332-5338

[31] Sharma R, Rawal RK, Malhotra M, et al. Design, synthesis and ex-vivo release studies of colon-specific polyphosphazene–anticancer drug conjugates. Bioorganic & Medicinal Chemistry. 2014;**22**:1104-1114

[32] Longley DB, Harkin DP, Johnston PG. 5-fluorouracil: Mechanisms of action and clinical strategies. Nature Reviews. Cancer. 2003;**3**:330-338

[33] Cameron DA, Gabra H, Leonard RC. Continuous 5-fluorouracil in the treatment of breast cancer. British Journal of Cancer. 1994;**70**:120-124

[34] Arias JL. Novel strategies to improve the anticancer action of 5-fluorouracil by using drug delivery systems. Molecules. 2008;**13**:2340-2369

[35] Zou M, Okamoto H, Cheng G, et al. Synthesis and properties of polysaccharide prodrugs of 5-aminosalicylic acid as potential colon-specific delivery systems. European Journal of Pharmaceutics and Biopharmaceutics. 2005;**59**:155-160

[36] Jung Y, Kim H-H, Kim H, et al. Evaluation of 5-aminosalicyltaurine as a colon-specific prodrug of 5-aminosalicylic acid for treatment of experimental colitis. European Journal of Pharmaceutical Sciences. 2006;**28**:26-33

[37] Varshosaz J, Emami J, Tavakoli N, et al. Synthesis and evaluation of dextran–budesonide conjugates as colon specific prodrugs for treatment of ulcerative colitis. International Journal of Pharmaceutics. 2009;**365**:69-76

[38] Hoste K, De Winne K, Schacht E. Polymeric prodrugs. International Journal of Pharmaceutics. 2004;**277**: 119-131

[39] Kumari S, Rath PK. Extraction and characterization of chitin and chitosan from (Labeo rohit) fish scales. Procedia Materials Science. 2014;**6**: 482-489

[40] Sun Z-J, Sun B, Sun C-W, et al. A poly(glycerol-sebacate-(5-fluorouracil-1-acetic acid)) polymer with potential use for cancer therapy. Journal of Bioactive and Compatible Polymers. 2012;**27**:18-30

[41] Malekshah RE, Shakeri F, Khaleghian A, et al. Developing a biopolymeric chitosan supported Schiff-base and Cu(II), Ni(II) and Zn(II) complexes and biological evaluation as pro-drug. International Journal of Biological Macromolecules. 2020;**152**: 846-861

[42] Barbosa HFG, Attjioui M, Ferreira APG, et al. New series of metal complexes by amphiphilic biopolymeric Schiff bases from modified chitosans: Preparation, characterization and effect of molecular weight on its biological applications. International Journal of Biological Macromolecules. 2020;**145**: 417-428

[43] Antony R, Arun T, Manickam STD. A review on applications of chitosan-based Schiff bases. International Journal of Biological Macromolecules. 2019;**129**: 615-633

[44] Aguirre-Rentería SA, Carrizales-Castillo JJJ, del Rayo Camacho Corona M, et al. Synthesis and in vitro evaluation of antimycobacterial and cytotoxic activity of new α,β-unsaturated amide, oxazoline and oxazole derivatives from l-serine.

Bioorganic & Medicinal Chemistry Letters. 2020;**30**:127074

[45] Hejazi L, Rezaee E, Tabatabai SA. Quinazoline-4(3H)-one derivatives as novel and potent inhibitors of soluble epoxide hydrolase: Design, synthesis and biological evaluation. Bioorganic Chemistry. 2020;**99**:103736

[46] Gao P, Wang L, Zhao L, et al. Anti-inflammatory quinoline alkaloids from the root bark of Dictamnus dasycarpus. Phytochemistry. 2020;**172**:112260

[47] Li B, Zhu F, He F, et al. Synthesis and biological evaluations of N′-substituted methylene-4-(quinoline-4-amino) benzoylhydrazides as potential anti-hepatoma agents. Bioorganic Chemistry. 2020;**96**:103592

[48] Das D, Xie L, Wang J, et al. In vivo efficacy studies of novel quinazoline derivatives as irreversible dual EGFR/HER2 inhibitors, in lung cancer xenografts (NCI-H1975) mice models. Bioorganic Chemistry. 2020;**99**:103790

[49] Huang ZH, Yin LQ, Guan LP, et al. Screening of chalcone analogs with anti-depressant, anti-inflammatory, analgesic, and COX-2-inhibiting effects. Bioorganic & Medicinal Chemistry Letters. 2020;**30**:127173

[50] Shao W-B, Zheng Y-T, Liu J-M, et al. Antibacterial activities against Ralstonia solanacearum and Xanthomonas oryzae pv. oryzae of 6-chloro-4-(4-substituted piperazinyl)quinazoline derivatives. Bioorganic & Medicinal Chemistry Letters. 2020;**30**:126912

[51] Ibrahim TS, Bokhtia RM, , et al. Design, synthesis and biological evaluation of novel 5-((substituted quinolin-3-yl/1-naphthyl) methylene)-3-substituted imidazolidin-2,4-dione as HIV-1 fusion inhibitors. Bioorganic Chemistry. 2020;**99**:103782

[52] Alafeefy AM, Kadi AA, Al-Deeb OA, et al. Synthesis, analgesic and anti-inflammatory evaluation of some novel quinazoline derivatives. European Journal of Medicinal Chemistry. 2010; **45**:4947-4952

[53] Mishra M, Agarwal S, Dixit A, et al. Integrated computational investigation to develop molecular design of quinazoline scaffold as promising inhibitors of plasmodium lactate dehydrogenase. Journal of Molecular Structure. 2020;**1207**:127808

[54] Wang M, Zhang G, Wang Y, et al. Design, synthesis and anti-influenza A virus activity of novel 2,4-disubstituted quinazoline derivatives. Bioorganic & Medicinal Chemistry Letters. 2020;**30**: 127143

[55] Bouzian Y, Karrouchi K, Sert Y, et al. Synthesis, spectroscopic characterization, crystal structure, DFT, molecular docking and in vitro antibacterial potential of novel quinoline derivatives. Journal of Molecular Structure. 2020;**1209**:127940

[56] Sureshkumar B, Mary YS, Panicker CY, et al. Quinoline derivatives as possible lead compounds for anti-malarial drugs: Spectroscopic, DFT and MD study. Arabian Journal of Chemistry. 2020;**13**:632-648

[57] Jiang Y, Fu C, Wu S, et al. Determination of the deacetylation degree of chitooligosaccharides. Marine Drugs. 2017;**15**:332

[58] Salama HE, Saad GR, Sabaa MW. Synthesis, characterization and biological activity of Schiff bases based on chitosan and arylpyrazole moiety.

International Journal of Biological Macromolecules. 2015;**79**:996-1003

[59] Synytsya A, Kim W-J, Kim S-M, et al. Structure and antitumour activity of fucoidan isolated from sporophyll of Korean brown seaweed Undaria pinnatifida. Carbohydrate Polymers. 2010;**81**:41-48

[60] Chen B, Xing J, Li M, et al. DOX@Ferumoxytol-medical chitosan as magnetic hydrogel therapeutic system for effective magnetic hyperthermia and chemotherapy in vitro. Colloids and Surfaces. B, Biointerfaces. 2020;**190**: 110896

[61] Wang Y, Cao H, Wang X. Synthesis and characterization of an injectable ε-polylysine/carboxymethyl chitosan hydrogel used in medical application. Materials Chemistry and Physics. 2020; **248**:122902

[62] Benltoufa S, Miled W, Trad M, et al. Chitosan hydrogel-coated cellulosic fabric for medical end-use: Antibacterial properties, basic mechanical and comfort properties. Carbohydrate Polymers. 2020;**227**:115352

[63] De Simone U, Spinillo A, Caloni F, et al. In vitro evaluation of magnetite nanoparticles in human mesenchymal stem cells: Comparison of different cytotoxicity assays. Toxicology Mechanisms and Methods. 2020;**30**:48-59

[64] Benedek TG. Methotrexate: From its introduction to non-oncologic therapeutics to anti-TNF-α. Clinical and Experimental Rheumatology. 2010;**28**: S3-S8

[65] Pinto JF. Site-specific drug delivery systems within the gastro-intestinal tract: From the mouth to the colon. International Journal of Pharmaceutics. 2010;**395**:44-52

[66] Shah K, Gupta JK, Chauhan NS, et al. Prodrugs of NSAIDs: A review. Open Medicinal Chemistry Journal. 2017;**11**: 146-195

[67] Liebert TF, Heinze T. Tailored cellulose esters: Synthesis and structure determination. Biomacromolecules. 2005;**6**:333-340

[68] Li D, Lu B, Zhang H, et al. Synthesis and in vitro evaluation of methotrexate conjugated O,N-carboxymethyl chitosan via peptidyl spacers. Journal of Nanoparticle Research. 2014;**16**:2609

[69] Ali SS, Kenawy ER, Sonbol FI, et al. Pharmaceutical potential of a novel chitosan derivative Schiff base with special reference to antibacterial, anti-biofilm, antioxidant, anti-inflammatory, hemocompatibility and cytotoxic activities. Pharmaceutical Research. 2019;**36**:5

[70] Gul S, Aziz-ur-Rehman AMA, et al. Synthesis, antimicrobial evaluation and hemolytic activity of 2-[[5-alkyl/aralkyl substituted-1,3,4-oxadiazol-2-yl]thio]-N-[4-(4-morpholinyl)phenyl]acetamide derivatives. Journal of Saudi Chemical Society. 2017;**21**:S425-S433

[71] Kumar SU, Gopinath P. Controlled delivery of bPEI-niclosamide complexes by PEO nanofibers and evaluation of its anti-neoplastic potentials. Colloids Surfaces B Biointerfaces. 2015;**131**: 170-181

Chapter 4

Chitosan Nanoparticle: Synthesis, Characterization, and Use as Plant Health Materials

Pranab Dutta, Arti Kumari and Madhusmita Mahanta

Abstract

Chitosan is a naturally occurring biopolymer having multifaceted applications in agriculture, medicine, food industry, and cosmetics. The association of this natural biopolymer with nanotechnology can produce revolutionary effects in plant protection and agriculture. Nano-chitosan can be fabricated using various methods. However, the green synthesis approach has gained attention in recent years. The green engineered nanoparticles are economical, energetically feasible, and environmentally benign. The biosynthesized nano-chitosan has evolved as a potential plant protection agent. Chitosan nanoparticles possess antifungal, antibacterial, and antiviral properties, and are found to be effective against seed-borne and soil-borne pathogens. Nano-chitosan also behaves as an effector molecule and induces local and systemic defense responses in plants. The mode of action of nano-chitosan involves alterations in membrane permeability, replication, cytoplasmic alterations, induction of defense-related genes, and cell lysis. Furthermore, chitosan nanoparticles can be used for soil improvement and can reduce pest and pathogen attacks, thereby promoting the growth of plants. The authors outline the methods of synthesis and characterization of chitosan nanoparticles, their utilization in plant protection and growth promotion, along with the underlying mechanisms.

Keywords: chitosan nanoparticles, biopolymer, green synthesis, characterization, plant protection

1. Introduction

Agriculture is a primary activity upon which the economic status of a country relies. The produce obtained from agricultural activity serves the purpose of mitigating domestic hunger as well as earning foreign currency. This demands the high production and productivity of crops. However, crop health, quality of production, and productivity are attributed to different biotic and abiotic factors associated with it. It has been already said that 20–40% of crop loss occurs around the globe due to attack of pest and diseases [1] that limits the yield of crops. The detrimental effect of chemical pesticides on Earth has resulted in the development of green strategies, such as sustainable farming, with the use of resistant varieties, biopesticides, nano bioformulations, integrated disease, and pest management.

As the interest of the scientific community shifted from chemocentric to sustainable agriculture, it opened up a vast possibility of exploiting nature-based biodegradable materials with potential biocontrol efficacy for plant disease management. With more intense research in this field led the scientists to speculate that the nano-sized materials may perform excellent activity as compared to the base source used for their synthesis. Therefore, the development of nano bioformulations and their use is encouraged to achieve the goal of sustainable farming. Different biopolymers viz., cellulose, starch, alginate, chitin, and chitosan, are used for the development of new materials with noble functionality and environmental sustainability. Among these, chitosan is the second most abundant biopolymer found in nature which is used widely due to its unique characteristics, such as abundance, large surface-to-volume ratio, biodegradability, biocompatibility, pH sensitivity, non-toxicity, and a safe alternative [2, 3]. Apart from that, nano-chitosan possesses antifungal and antibacterial activity along with other plant growth-promoting traits [4, 5] which makes it promising in several aspects of plant growth and development.

This chapter aims to briefly review the importance, green synthesis, characterization, mode of action, and successful use of chitosan nanoparticles (ChNPs) for the effective management of plant diseases.

2. Importance of chitosan nanoparticles

Chitin and chitosan are the primary components of crustacean shells, such as shrimp, squid, crab, and lobster, the exoskeleton of terrestrial insects viz., honeybees and silkworm, and cell walls of fungi like molds, yeast, and ray fungi, such as *Streptomyces* [6, 7]. Chitosan is a cationic biopolymer obtained by the whole or partial deacetylation of chitin. It is a linear polysaccharide consisting of (1–4)-linked 2-amino-2-deoxy-β-D-glucopyranose obtained after deacetylation of N-acetyl-D-glucosamine [8, 9]. The term chitosan does not specifically indicate a unique compound but a group of co-polymers owing to their degree of deacetylation, polymerization, molecular mass, viscosity, and acid dissociation constant, that is, pka [10]. The chitosan derived from microbial sources is considered as promising as the process underlying can be manipulated to prepare a pure and uniform product with desired specific characteristics [11]. It is a very versatile biopolymer having multifaceted activity in the field of medicine, agriculture, food industry, cosmetics, and sewage treatment (**Figure 1**) [12]. ChNPs are widely used due to their unique polymeric cationic character, absorption enhancing effects, mucoadhesive nature, biocompatibility, and biodegradability. Being a modified linear polysaccharide with varying numbers of free amino groups in their polymeric chain with cationic property, chitosan offers ionic cross-linking of multivalent anions, which is making it a significant biopolymer for the synthesis of nanoparticles [6, 13]. The positively charged ChNPs have more affinity toward the negatively charged biological membrane and site-specific targeting *in vivo* [14]. Chitosan when applied to foliage or soil, can elicit innate defense response within the plant to resist insect and pathogen attack [5] by the production of antifungal hydrolases, phytoalexins, or by inducing structural barriers via the synthesis of lignin-like material [15, 16]. Further, ChNPs significantly have a positive impact on the biophysical properties of the plant. The application of nano-chitosan increases the rate of photosynthesis, induces root nodulation, upregulates nutrient uptake, enhances the rate of germination of seed, and boosts plant vigor [5, 17]. It is widely used in the delivery of fertilizers, micro-nutrients, and pesticides as it ensures

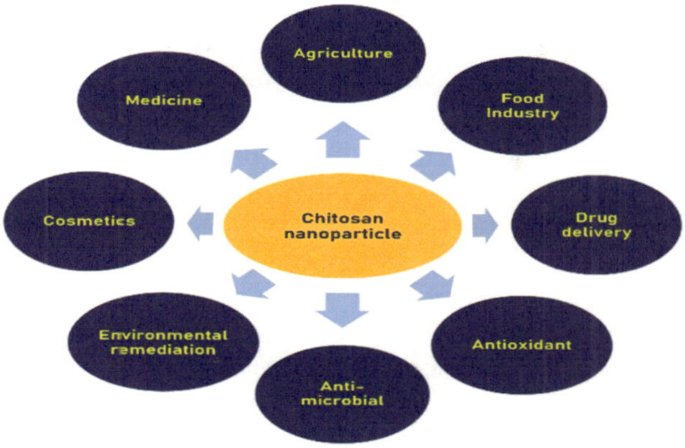

Figure 1.
Applications of nano-chitosan.

slow release of the drugs with enhanced solubility [10]. Moreover, chitosan-mediated genetic transformation is also a successful one as it forms a complex via electrostatic interaction and protects the nucleic acid from nuclease degradation. It gives rise to stably transformed plants as compared to those developed via traditional methods of gene delivery [18, 19].

3. Green synthesis of chitosan nanoparticles

Green synthesis is a novel method of synthesis of NPs using microorganisms, such as bacteria, fungi, actinomycetes, and botanicals [20–25]. It is a bottom-up approach to nanoparticle synthesis which interconnects two disciplines of science, that is, nanotechnology and biotechnology. Biogenic synthesis of nanoparticles is preferred over chemical or physical synthesis methods as it is a safe, environmentally benign, economically and energetically feasible, less time-consuming process, and it makes optimum use of the redox potential of metabolites produced by biological entities to convert the macromolecules to nano form [9, 21, 26–30]. The main principle of the biogenic synthesis of the nanoparticle is based on the redox reactions that occur when the microorganisms/biological entity grabs the metal ion and detoxifies it to element metal through the enzymatic activity of the cell. It can be categorized into intracellular and extracellular synthesis. In intracellular synthesis, the metal ions are transported into the cell and the reduction reaction occurs within the cell cytoplasm, cell wall, and/or periplasmatic space, therefore, the resultant nanoparticles form inside the cell. However, the latter involves the synthesis of nanoparticles on the cell surface via the catalytic activity of reductase enzyme upon the trapped foreign entity [31]. Harvesting of nanoparticles from the cell matrix is a tedious process in intracellular synthesis which is why extracellular synthesis of nanoparticles is preferred mostly. The microbes are often regarded as eco-friendly green nano-factories due to their large-scale production ability of nanoparticles with relative control over their size and shape (regulated by surrounding environmental conditions) with a simpler process of production [26].

In phytofabrication of nanoparticles, generally, the metal ions are added to the plant part extract and then continuously stirred in a magnetic stirrer. The formation of nanoparticles can be confirmed visually when the resultant solution changes its color [30, 32]. It is due to the activity of the antioxidant, such as polyphenol, flavonoids, and phytoalexins, that act upon the metal ion and convert it to nontoxic element metal in nanoform. Therefore, a plant with a higher percentage of natural reducing constituents is desirable for phytofabrication of nanoparticles. *Pelargonium graveolens* L'Her commonly known as the rose geranium plant is a rich source of natural antioxidants. Its leaves and essential oil have several therapeutic applications in the field of pharmacology [33, 34]. El-Naggar et al. [9] mixed an equal volume of chitosan solution and phytoextract of *P. graveolens* and incubated the mixture at 50°C in a rotary shaker. The resultant turbid solution indicating the ongoing redox reaction is then centrifuged at 10,000 × g for 10 min. It was further washed with an acetic acid solution to remove the unreacted chitosan and the ChNPs present in the solution are extracted by subjecting it to freeze drying. Similarly, Boruah and Dutta [5] biogenically synthesized ChNPs from fungal sources rich in chitosan. Isolation of chitosan was done by treating the fungal biomass with a series of alkali and acid treatments under controlled conditions. Initial treatment of fungal biomass with NaOH yields an alkali-insoluble material (AIM) that was further subjected to an acid treatment to extract the fungal chitosan. It was then converted into nano-chitosan by adding 1% TPP solution in a magnetic stirrer.

Guzman et al. [35] synthesized colloidal AgNps by a combination of ultrasonication and chemical reduction methods using gallic acid and chitosan, respectively, and during characterization, they found that as-synthesized gallic acid-chitosan modified silver nanoparticles (GC-AgNps) were monodispersed, spherical shape with an average size of 26.23 ± 9.92 nm, and stable for four weeks without any noticeable change in size. GC-AgNps were found highly effective against *Escherichia coli* even at 1 µg/mL after 120 min of exposure.

4. Characterization of chitosan nanoparticles

The nanoparticle formation is greatly affected by the processing conditions and time. They are characterized on the basis of their surface plasmon resonance, morphology, particle size distribution, zeta potential, functional group analysis, etc., using UV-vis spectrophotometer, electron microscope, dynamic light scattering, Fourier transform infrared (FTIR) spectroscopy [36]. Moreover, atomic absorption spectroscopy is done to study the release profile of Ch-encapsulated nanoparticles.

4.1 Characterization by UV-vis spectroscopy

UV-vis spectroscopy is the confirmatory analysis that ascertains the formation of nanoparticles by surface plasmon resonance. UV-vis absorption spectroscopy is used to examine the optical properties of ChNPs obtained from the commercial production of chitosan showed an absorption band at 330.25 nm, whereas those obtained from biogenic sources exhibited the absorption band at 310–342 nm [5]. Kain and Kumar [37] reported a band at 200–300 nm for the biogenically synthesized ChNPs from common yarrow. Moreover, AbdElhady [38] studied the UV-vis spectrophotometer reading of ChNPs prepared at two different temperatures, that is, 40°C, 60°C, and 80°C, and found that the absorption band was obtained at 356, 348, and 353 nm, respectively.

4.2 Characterization by dynamic light scattering

The main principle of DLS is based on the Brownian movement of particles/molecules present in the solution that results from their collision with the randomly moving solvent particles. A laser beam is passed through the sample, and the fluctuation in scattered light due to the random motion of particles is detected by the photon detector. DLS is used for the measurement of average particle size, particle size distribution, polydispersity index (PDI), and zeta potential. PDI explains the polydispersity or monodispersity of particles in an aqueous medium. PDI value greater than 0.5 represents polydispersity and less than 0.5 normally shows the monodispersity of particles. Generally, monodisperse ChNPs exhibit the PDI value within the range of 0.2–0.4 [2]. Further, the surface charge of nanoparticles, also known as zeta potential, explains the stability of the nanoparticles, which is measured in the range of ±30 mV. The ChNPs show a positive zeta potential value that may vary from 11.2 ± 1.2 mV to 18.7 ± 0.4 mV [5, 39]. Further, the appropriate particle size determined by DLS showed that the ChNPs are nearly spherical in shape with size ranging from 150 to 350 nm [40]. However, Sivakami et al. [41] prepared ChNPs by cross-linking low molecular weighed chitosan with TPP and found a minimum particle size <100 nm. Similarly, Boruah and Dutta [5] reported that the biogenically synthesized ChNPs exhibited their size within the range from 78.36 to 300.1 nm. A mean particle size of 50 nm was obtained by Sahab et al. [42] for the chitosan poly acrylic acid nanoparticles.

4.3 Characterization by electron microscopy

The internal and external morphology of the nanoparticles is studied using a transmission electron microscope and scanning electron microscope. Transmission electron microscopy revealed that the ChNPs are often spherical shaped with an amorphous nature [5, 14, 43]. Similarly, scanning electron microscopy studies conducted by many researchers found that the ChNPs are nearly spherical shaped with the smooth external surface [44, 45]. Parida et al. [46] obtained round ChNPs with a 78 nm diameter. Similarly, Kain and Kumar [37] found that the green synthesized ChNPs of common yarrow (*Achillea millefolium*) are smooth surfaced, spherical with a diameter less than 100 nm with a smallest diameter of 4.15 nm.

4.4 Characterization by Fourier transform infrared spectroscopy

FTIR study is conducted to confirm the synthesis of nanoparticles by determining their functional groups. Sample preparation for FTIR is done by gently triturating it with KBr which is then compressed into disks. The compressed disks are scanned against a blank KBr pellet background at 25°C to obtain the FTIR results. For every spectrum, a 32 scan interferogram was collected at transmittance/absorbance mode in the 4000–400 cm^{-1} region [2]. The functional groups of a chitosan nanoparticle consist of amide ($-NH_2$) and hydroxyl ($-OH$) group, C$-$H, C$-$N, C$-$O, and P=O stretching [5]. Generally, the FTIR peak at 3000–3500 cm^{-1} attributed to ($-OH$) and ($-NH_2$) is the confirmatory peak for the formation of ChNPs [46]. Sharma et al. [45] obtained a wider peak of the hydroxyl group (3200–3600), which led them to conclude that hydrogen bonding is enhanced in ChNPs when analyzed by FTIR. Choudhary et al. [2] found a band at 3424 cm^{-1} for the synthesized chitosan nanoparticles that represent the stretching vibration of the combined peaks of the amide

(−NH_2) and hydroxyl (−OH) group. Kain and Kumar [37] reported FTIR peaks at 3317.48, 2139.29, and 1638.46 as a confirmation of the formation of ChNPs. Similarly, Boruah and Dutta [5] obtained a strong and broadband at 3250 cm^{-1} signifying the stretching between hydroxyl and amide groups. They have also found other peaks at 2865 cm^{-1}, 1182 cm^{-1}, 1642 cm^{-1}, and 1182 cm^{-1} is attributed to C−H stretching, asymmetric C−O stretching, stretching between C=O and N−H banding, and P=O.

5. Application of chitosan nanoparticles in plant health management

ChNPs have emerged as a potential antimicrobial agent and found effective against numerous phytopathogens. Nano-chitosan is reported to be effective against seed-borne as well as against soil-borne pathogens. It behaves as an elicitor of plant defense responses, inducing both local and systemic defense responses. Thus, nano-chitosan can be used as a plant health material by protecting the plant from biotic and abiotic stresses and by promoting the growth of the plants.

ChNPs possess a greater affinity toward the membrane of microorganisms and can easily penetrate the pathogen's cell [47]. The smaller size and greater surface area of ChNPs increase the antimicrobial efficiency of chitosan biopolymer. Several studies confirmed ChNPs as an effective plant health management agent due to their dual role as plant protection and plant growth stimulating agents.

ChNPs have been known to possess antifungal properties against numerous phytopathogenic fungi. Nano-chitosan exhibit greater efficacy as compared to its bulk counterpart as NPs can negotiate cell wall and cell membranes more effectively due to their unique physico-chemical properties. It can effectively inhibit the development of phytopathogenic fungi at any stage of their life cycle. Chitosan can completely inhibit spore germination, germ tube elongation, and mycelial growth of fungi [48], and it can penetrate the cell membrane by plasma membrane permeabilization and results in cell lysis [49]. Boruah and Dutta [5] biogenically synthesized chitosan nanoparticles using four different fungal sources *viz*., *Beauveria bassiana, Fusarium oxysporum, Trichoderma viride,* and *Metarhizium anisopliae. In vitro* assay suggested that synthesized ChNPs in combination with *T. asperellum* was effective in suppressing mycelial growth of soil-borne fungal pathogens *viz*., *Rhizoctonia solani, Fusarium oxysporum,* and *Sclerotium rolfsii*. Abdel-Rahman et al. [50] studied the efficacy of chitosan (Ch) (2 and 4 g/L) as well as ChNPs (0.2 and 0.4 g/L) against blue rot disease of apples caused by *Penicillium expansum*. They observed that ChNPs performed better than compared to their bulk counterpart for both natural and artificial infections. Also, the fruit quality parameters, such as firmness, titratable acidity, and total soluble solids, were kept intact. The expression of defense-related genes *viz*., chitinase, β-1,3-glucanase, peroxidase, phenylalanine ammonia lyase-1 (PAL1), xyloglucan endotransglycosylase (XET), and pathogenesis-related protein (PR8), were also upregulated indicating the development of systemic acquired resistance in plants against the pathogen. Saharan et al. [51] synthesized ChNPs using the ionic gelation method and its efficacy was determined against phytopathogenic fungi *viz*., *Macrophomina phaseolina*, *Alternaria alternata,* and *Rhizoctonia solani*. They observed a decline in the radial growth of the fungi in a dose-dependent manner. Muthukrishnan and Ramalingam [52] synthesized ChNPs biogenically mediating *Penicillium oxalicum*. The nanomaterial was found effective against *Fusarium oxysporum ciceri*, *Pyricularia grisea,* and *Alternaria solani* with the rate of inhibition 87%, 92%, and 72%, respectively. Also, seed treatment with ChNPs exhibited positive morphological effects, such as enhanced germination percentage, seed vigor index, and biomass content in chickpeas. The efficacy of nanomaterials depends on their size, charge, and permeability through biological membranes. Again, the *in vivo* assay conducted under detached leaf condition observed 100% suppression of blast disease symptoms when treated with ChNPs prepared using the ionic gelation method [53]. Kheiri et al. [54] synthesized ChNPs from chitosans of different molecular weight and observed their antifungal activity against *Fusarium graminearum* causing fusarium head blight in wheat. The dynamic light scattering analysis showed a variable size of synthesized nanomaterials (180.9, 339.4, 225.7, and 595.7 nm). The inhibitory effect of these NPs was tested at different concentrations and maximum mycelial growth reduction (77.5%) was observed at 5000 ppm. The results obtained from greenhouse trials indicated a decline in the area under the disease progress curve (AUDPC) in NP-treated plants.

The reports of ChNPs as an antibacterial agent against plant pathogenic bacteria is very scarce and need further thrust in this domain. ChNPs were found effective against *Ralstonia solanacearum* causing bacterial wilt of tomato and potato. *In vitro* experiment indicated an increase in the inhibition zone with increasing concentration of nano-chitosan and found highest at 200 μg/ml concentration. *In vivo* assay revealed foliar application of nano-chitosan led to a decline in the disease incidence and severity in bacterial wilt-infected tomato and potato plants [55]. ChNPs directly interact with the bacterial cell wall and may cause modification in the external shape, loss of flagella, and lysis of the cell. The RAPD-PCR results showed differences in the genotype of treated *Ralstonia solanancearum* as compared to the genotype of untreated isolates [55]. The antibacterial activity of Cu-chitosan nanoparticles against *Pseudomonas syringae* pv. *glycinea* causing bacterial blight of soybean was reported by Choudhary et al. [56]. Concentration of 1000 ppm was found most effective in controlling the bacterial pathogen. An *in vitro* assessment was conducted with ChNPs and chitosan nanocomposites with lime essential oil and thyme critical oil against *Pectobacterium carotovorum*. The results indicated that chitosan nanocomposites with thyme essential oil were effective in producing an inhibition zone [57]. ChNPs were also observed to possess potentially high antibacterial activity against *P. fluorescens* and *Erwinia carotovora* causing bacterial soft rot [58]. Oh et al. [8] reported the antibacterial activity of ChNPs against phytopathogenic bacteria *viz., E. carotovora* subsp. *carotovora* and *X. campestris* pv. *vesicatoria*. Santiago et al. [59] biogenically synthesized Ch-derived NPs containing AgNPs and observed its antibacterial efficacy against *R. solanacearum* causing bacterial wilt of tomato. They found Ag-NP entrapped chitosan as a suitable alternative to chemical bactericides. Cs/TiO_2NPs were found effective against the most dreaded bacterial pathogen of rice *viz.*, *X. oryzae* pv. *oryzae* [60].

The chitosan biopolymer was found to inhibit the systemic propagation of viruses and virus-like organisms in infected plants and induce a host hypersensitive response against the viral pathogen [61–63]. The molecular weight of chitosan affects the degree of suppression of viral infections [64]. However, none of the studies has practically proved the ability of the chitosan molecule to absolutely inactivate the virus particles. Most of the studies have reported the inactivation of viral replication that prevents the multiplication and subsequent spread of the virus particles systemically. It may be hypothesized that ChNPs having a smaller size and greater surface area can easily penetrate into host tissues and tightly binds with nucleic acid causing selective inhibition and ramification of virus particles. Lu et al. [65] reported inhibition of TMV in tobacco pants when oligochitosan (50 μg ml^{-1}) was applied 24 h before inoculation. The epidermal cells of tobacco leave treated with oligochitosan showed an increase in

Chitosan nanomaterial	Application	Target pathogen	Effect	Reference
Chitosan nanocomposite with *T. asperellum*	Antifungal	*Rhizoctonia solani, Fusarium oxysporum* and *Sclerotium rolfsii.*	Reduction in mycelial growth	[5]
Chitosan NPs	Antifungal	*A. alternata, M. phaseolina* and *R. solani*	Reduction in mycelial growth	[51]
Chitosan	Antifungal	*Alternaria kikuchiana* Tanaka and *Physalospora piricola*	Inhibition of spore germination, germ tube elongation, and mycelial growth	[48]
Rhodamine-labeled chitosan	Antifungal	*F. oxysporum*	Inhibition of mycelial growth, cell membrane permealization, and lysis	[49]
Chitosan NPs	Antifungal and plant growth promotion	*Penicillium expansum*	Reduction in natural and artificial infections. Induction of defense-related genes and resistance in plants	[50]
Chitosan NPs	Antifungal and plant growth promotion	*Sclerospora graminicola*	Promote seed germination and seedling vigor, induced systemic and durable resistance	[66]
Ag-chitosan nanocomposites	Antifungal	Seed-borne fungi	Inhibition of mycelial growth	[67]
Cu-chitosan nanocomposites	Antifungal	*Fusarium graminearum*	Inhibition of mycelial growth	[68]
Chitosan nanocomposites	Antibacterial	*Ralstonia solanacearum*	Inhibition zone production, cell wall modification, loss of flagella and cell lysis	[55]
Chitosan nanocomposites with thyme critical oil	Antibacterial	*Pectobacterium carotovorum*	Inhibition zone production	[57]
Cs/TiO2NPs	Antibacterial	*X. oryzae* pv. *oryzae*	Reduction in infection	[60]

Chitosan nanomaterial	Application	Target pathogen	Effect	Reference
Oligochitosan	Induction of resistance in plants against TMV	*Tobacco Mosaic Virus*	Increase in the levels of intracellular H_2O_2, NO, and phenylalanine ammonia-lyase (PAL)	[65]

Table 1.
Antimicrobial properties of chitosan-based nanomaterials.

the levels of intracellular H_2O_2, NO, and increased activity of phenylalanine ammonia-lyase (PAL) indicating induction of plant defense response against TMV (**Table 1**).

6. Mode of action of chitosan nanoparticles

6.1 Direct activity against pathogen

ChNPs can directly interact with the cellular membrane of microorganisms due to their unique physicochemical properties and can easily permeate into the cytoplasm. The direct mode of action of ChNPs against fungi includes inhibition of spore germination, germ tube elongation, mycelial growth, and cell lysis. Benhamou [69] conducted ultrastructural studies and reported that chitosan induces numerous structural and morphological changes leading to distorted hyphae. This can be explained as the chitosan particles are polycationic in nature, it allows alteration in membrane permeability and cytoplasmic aggregations. As a result, the activity of enzymes involved in the synthesis and assembly of cell wall polymers are dwindled. The antibacterial effect includes disruption of the bacterial cell wall, cellular membrane, loss of external appendages, such as flagella, finally, leading to cell lysis. None of the studies have proved ChNPs to inactivate viruses and viroids. Most of the studies reported inhibition of virus replication, multiplication, and spread by chitosan. However, against pests and pathogens, ChNPs operate via an indirect mechanism, such as induction of host resistance.

6.2 Indirect mechanism

Chitosan molecule is generally used as an elicitor rather than an antimicrobial agent in plant disease control. It can be recognized by the plant PRRs and can trigger a cascade of defense responses. Chitosan molecule behaves as MAMP/PAMP or general elicitor and induces nonhost resistance in plants along with priming systemic immunity [70]. The cascade of biochemical and molecular reactions induced by chitosan includes enhanced H^+ and Ca^{2+} influx into the cytosol, callose apposition, activation of MAP-kinases, hypersensitive response, oxidative burst, synthesis of phytohormones *viz.*, jasmonates and abscisic acid, as well as phytoalexins and PR-proteins [71].

6.3 Physical barrier in pathogen penetration

ChNPs can agglutinate around the penetration sites of the pathogen after its application on plant tissues and has two major effects. The first effect includes

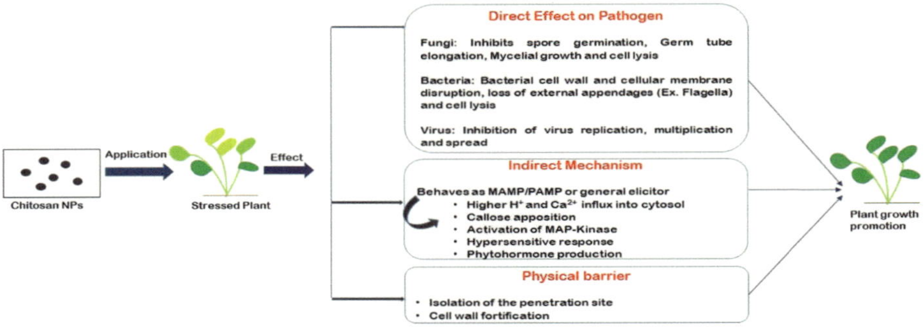

Figure 2.
Schematic representation of the mechanism of action of chitosan nanoparticles on plants.

isolation of the penetration site from healthy tissues by forming a physical barrier that prevents further spread of the pathogen. Around the isolated zone, several biochemical changes occur that lead to the elicitation of hypersensitive response, accumulation of H_2O_2 and other free radicals, which lead to cell wall fortification and induction of systemic acquired resistance. The second effect includes the initiation of wound healing process by binding with various materials (**Figure 2**).

6.4 Plant growth promotion

Nano-chitosan imparts an eustress effect on seedling germination and plant growth parameters, such as plant height, shoot length, root length, and biomass content, which have been confirmed through a series of studies. The study conducted on the effect of nano-chitosan on *Phaseolus vulgaris* L. under salt stress conditions revealed that 0.3% nano-chitosan was the best treatment in terms of germination, growth parameters. Also, significant increase in M.S.I, Chl.a, Chl.b, proline, catalase, carotenoids, and antioxidant enzymes were observed [72].

7. Conclusion and future prospects

Chitosan is a naturally occurring miracle compound having enthralling antimicrobial and eliciting properties. Nano-chitosan is gaining attention nowadays due to its greater efficacy and biosafety. Chitosan nanomaterials can be used in varied ways for plant disease management, thereby preserving crop quality and yield. In recent years, several findings have been gathered indicating nano-chitosan as a potential plant health material. However, more studies need to be channelized to unveil the exact mode of action of nano-chitosan specific to the pathosystem. Incorporation of chitosan nanomaterials into integrated pest management practices by devising suitable incorporation techniques need to be pursued. The biopolymer-based nanomaterials need extensive exploration owing to their multifunctional properties and diverse mechanisms. In the coming years, the use of nano-chitosan for combating biotic and abiotic stresses and transport of agrochemicals would be a promising discipline for utility in sustainable agriculture.

Author details

Pranab Dutta*, Arti Kumari and Madhusmita Mahanta
School of Crop Protection, College of Post Graduate Studies in Agricultural Sciences, Central Agricultural University (Imphal), Umiam, Meghalaya, India

*Address all correspondence to: pranabdutta74@gmail.com

References

[1] CABI. Global burden of crop loss. 2019. Available from: https://www.cabi.org/what-we-do/cabi-projects/

[2] Choudhary RC, Kumaraswamy RV, Kumari S, Pal A, Raliya R, Biswas P, et al. Synthesis, characterization, and application of chitosan nanomaterials loaded with zinc and copper for plant growth and protection. In: Prasad R, Kumar M, Kumar V, editors. Nanotechnology: An Agricultural Paradigm. Singapore: Springer; 2017. pp. 227-247

[3] Ali A, Ahmed S. A review on chitosan and its nanocomposites in drug delivery. International Journal of Biological Macromolecules. 2018;**109**:273-286

[4] Saharan V, Pal A. Chitosan Based Nanomaterials in Plant Growth and Protection. New Delhi: Springer; 2016. pp. 33-41

[5] Boruah S, Dutta P. Fungus mediated biogenic synthesis and characterization of chitosan nanoparticles and its combine effect with *Trichoderma asperellum* against *Fusarium oxysporum*, *Sclerotium rolfsii* and *Rhizoctonia solani*. Indian Phytopathology. 2021;**74**(1):81-93. DOI: 10.1007/s42360-020-00289-w

[6] Perera UMSP, Rajapakse N. Chitosan nanoparticles: Preparation, characterization, and applications. In: Seafood Processing By-products. New York: Springer; 2014. pp. 371-387

[7] Sivashankari PR, Prabaharan M. Deacetylation modification techniques of chitin and chitosan. In: Chitosan Based Biomaterials. Vol. 1. India: Woodhead Publishing; 2017. pp. 117-133

[8] Oh JW, Chun SC, Chandrasekaran M. Preparation and in vitro characterization of chitosan nanoparticles and their broad-spectrum antifungal action compared to antibacterial activities against phytopathogens of tomato. Agronomy. 2019;*9*(1):21. DOI: 10.3390/agronomy9010021

[9] El-Naggar NEA, Saber WI, Zweil AM, Bashir SI. An innovative green synthesis approach of chitosan nanoparticles and their inhibitory activity against phytopathogenic *Botrytis cinerea* on strawberry leaves. Scientific Reports. 2022;**12**(1):1-20

[10] Malerba M, Cerana R. Chitosan effects on plant systems. International Journal of Molecular Sciences. 2016;**17**(7):996

[11] Knežević-Jugović Z, Petronijević Ž, Šmelcerović A. Chitin and chitosan from microorganisms. In: Chitin, Chitosan, Oligosaccharides and Their Derivatives: Biological Activities and Applications Boca Raton: CRC Press; 2020. pp. 25-34

[12] Yanat M, Schroën K. Preparation methods and applications of chitosan nanoparticles; with an outlook toward reinforcement of biodegradable packaging. Reactive and Functional Polymers. 2021;**161**:104849

[13] Agnihotri SA, Mallikarjuna NN, Aminabhavi TM. Recent advances on chitosan-based micro-and nanoparticles in drug delivery. Journal of Controlled Release. 2004;**100**(1):5-28

[14] Qi L, Xu Z, Jiang X, Hu C, Zou X. Preparation and antibacterial activity of chitosan nanoparticles. Carbohydrate Research. 2004;**339**(16):2693-2700

[15] Hernandez-Lauzardo AN, Bautista-Baños S, Velazquez-Del

Valle MG, Méndez-Montealvo MG, Sánchez-Rivera MM, Bello-Perez LA. Antifungal effects of chitosan with different molecular weights on in vitro development of *Rhizopus stolonifer* (Ehrenb.: Fr.) Vuill. Carbohydrate Polymers. 2008;**73**(4):541-547

[16] Anusuya S, Sathiyabama M. Effect of Chitosan on Rhizome Rot Disease of Turmeric Caused by *Pythium aphanidermatum*. International Scholarly Research Notices; 2014

[17] Van SN, Minh HD, Anh DN. Study on chitosan nanoparticles on biophysical characteristics and growth of Robusta coffee in green house. Biocatalysis and Agricultural Biotechnology. 2013;**2**(4):289-294

[18] Mao S, Sun W, Kissel T. Chitosan-based formulations for delivery of DNA and siRNA. Advanced Drug Delivery Reviews. 2010;**62**(1):12-27

[19] Iriti M, Varoni EM. Chitosan-induced antiviral activity and innate immunity in plants. Environmental Science and Pollution Research. 2015;**22**(4):2935-2944

[20] Das A, Dutta P. Antifungal activity of biogenically synthesized silver and gold nanoparticles against sheath blight of rice. Journal of Nanoscience and Nanotechnology. 2021;**21**(6):3547-3555

[21] Dutta P, Kaman P, Kaushik H, Boruah S. Biotechnological and nanotechnological approaches for better plant health management. Trends in Biosciences. 2015;**8**(22):6051-6065

[22] Kaman PK, Dutta P. Nanocentric plant health management with special reference to silver. International Journal of Current Microbiology and Applied Sciences. 2017;**6**(6):2821-2830

[23] Kaman PK, Boruah S, Kaushik H, Dutta P. Effect of biosynthesized silver nanoparticles on morphophysiology of host. International Journal of Botany and Research. 2018;**8**(3):1-4

[24] Goswami R, Bhattacharyya A, Dutta P. Nanotechnological approach for management of anthracnose and crown rot diseases of banana. Journal of Mycology and Plant Pathology. 2020;**50**(4):370-381

[25] Ahmed AA, Dutta P. Effect of green synthesized silver nanoparticles on soil properties. Journal of Medicinal Plant Studies. 2020;**8**(1):2320-3862

[26] Li X, Xu H, Chen ZS, Chen G. Biosynthesis of nanoparticles by microorganisms and their applications. Journal of Nanomaterials. 2011

[27] Gogate S, Rahman S, Dutta P. Efficacy of synthesized nanoparticles using *Ocimum sanctum* (L.) leaf extract against *Corcyra cephalonia* (S.). Journal of Entomology and Zoology Studies. 2018;**6**(3):1149-1155

[28] Gogate S, Rahman S, Dutta P. Pesticidal activity of green synthesized nanoparticles using *Nerium olender* (L.) leaves extract against *Leucinodes arbonalis* (G.). International Journal of Chemical Studies. 2018;**6**(3):438-442

[29] Bhuyan B, Paul A, Paul B, Dhar SS, Dutta P. *Paederia foetida* Linn. promoted biogenic gold and silver nanoparticles: Synthesis, characterization, photocatalytic and in vitro efficacy against clinically isolated pathogens. Journal of Photochemistry and Photobiology B: Biology. 2017;**173**:210-215

[30] Sharma A, Dutta P, Mahanta M, Kumari A, Yasin A. Botanicals as a source of nanomaterial for pest and disease management. Plant Health Archives. 2022;**1**(1):5-9

[31] Mohanpuria P, Rana NK, Yadav SK. Biosynthesis of nanoparticles: Technological concepts and future applications. Journal of Nanoparticle Research. 2008;**10**(3):507-517

[32] Dutta P, Kaman PK. Nanocentric plant health management with special reference to Silver. International Journal of Current Microbiology and Applied Sciences. 2017;**6**(6):2812-2830

[33] Ćavar S, Maksimović M. Antioxidant activity of essential oil and aqueous extract of Pelargonium graveolens L'Her. Food Control. 2012;**23**(1):263-267

[34] Boukhris M, Bouaziz M, Feki I, Jemai H, El Feki A, Sayadi S. Hypoglycemic and antioxidant effects of leaf essential oil of *Pelargonium graveolens* L'Hér. in alloxan induced diabetic rats. Lipids in Health and Disease. 2012;**11**(1):1-10

[35] Guzmán K, Kumar B, Vallejo MJ, Grijalva M, Debut A, Cumbal L. Ultrasound-assisted synthesis and antibacterial activity of gallic acid-chitosan modified silver nanoparticles. Progress in Organic Coatings. 2019;**129**:229-235

[36] Kaman P, Dutta P. Synthesis, characterization and antifungal activity of biosynthesized nanoparticle. Indian Phytopathology. 2019;**72**:79-88

[37] Kain D, Kumar S. Synthesis and characterization of chitosan nanoparticles of *Achillea millefolium* L. and their activities. F1000 Research. 2020;**9**:1297

[38] AbdElhady MM. Preparation and characterization of chitosan/zinc oxide nanoparticles for imparting antimicrobial and UV protection to cotton fabric. International Journal of Carbohydrate Chemistry. 2012

[39] Nagarajan E, Shanmugasundaram P, Ravichandiran V, Vijayalakshmi A, Senthilnathan B, Masilamani K. Development and evaluation of chitosan based polymeric nanoparticles of an antiulcer drug lansoprazole. Journal of Applied Pharmaceutical Science. 2015;**5**:20-25

[40] Mohammadpour DN, Eskandari R, Avadi MR, Zolfagharian H, Mohammad Sadeghi A, Rezayat M. Preparation and in vitro characterization of chitosan nanoparticles containing Mesobuthuseupeus scorpion venom as an antigen delivery system. Journal of Venomous Animals and Toxins Including Tropical Diseases. 2012;**18**(1):44-52

[41] Sivakami MS, Gomathi T, Venkatesan J, Jeong H, Kim S, Sudha PN. Preparation and characterization of nano chitosan for treatment waste water. International Journal of Biological Macromolecules. 2013;**57**:204-212

[42] Sahab AF, Waly AI, Sabbour MM, Nawar LS. Synthesis, antifungal and insecticidal potential of Chitosan (CS)-g-poly (acrylic acid) (PAA) nanoparticles against some seed borne fungi and insects of soybean. International Journal Chemtech Research. 2015;**8**(2):589-598

[43] Vaezifar S, Razavi S, Golozar MA, Karbasi S, Morshed M, Kamali M. Effects of some parameters on particle size distribution of chitosan nanoparticles prepared by ionic gelation method. Journal of Cluster Science. 2013;**24**(3):891-903

[44] Tao Y, Zhang H, Gao B, Guo J, Hu Y, Su Z. Water-Soluble chitosan nanoparticles inhibit hypercholesterolemia induced by feeding a high-fat diet in male Sprague-Dawley rats. Journal of Nanomaterials. 2011;**2011**:1-5. DOI: 10.1155/2011/814606

[45] Sharma K, Somavarapu S, Colombani A, Govind N, Taylor KM. Nebulised siRNA encapsulated crosslinked chitosan nanoparticles for pulmonary delivery. International Journal of Pharmaceutics. 2013;**455**(1-2):241-247

[46] Parida UK, Rout N, Bindhani BK. In vitro properties of chitosan nanoparticles induce apoptosis in human lymphoma SUDHL-4 cell line. Advances in Bioscience and Biotechnology. 2013;**4**:1118-1127

[47] Van CN, Van BN, Ming-Fa H. Curcumin-loaded chitosan/gelatin composite sponge for wound healing application. International Journal of Polymer Science. 2013;**2013**:106570

[48] Meng XH, Yang LY, Kennedy JF, Tian SP. Effects of chitosan and oligochitosan on growth of two fungal pathogens and physiological properties in pear fruit. Carbohydrate Polymers. 2010;**81**:70-75

[49] Palma-Guerrero J, López-Jiménez JA, Pérez-Berná AJ, Huang IC, Jansson HB, Salinas J, et al. Membrane fluidity determines sensitivity of filamentous fungi to chitosan. Molecular Microbiology. 2010;**75**:1021-1032

[50] Abdel-Rahman FA, Monir GA, Hassan MSS, Ahmed Y, Refaat MH, Ismail IA, et al. Exogenously applied chitosan and chitosan nanoparticles improved apple fruit resistance to Blue Mold, upregulated defense-related genes expression, and maintained fruit quality. Horticulturae. 2021;**7**(8):224

[51] Saharan V, Mehrotra A, Khatik R, Rawal P, Sharma SS, Pal A. Synthesis of chitosan-based nanoparticles and their in vitro evaluation against phytopathogenic fungi. International Journal of Biological Macromolecules. 2013;**62**:677-683

[52] Muthukrishnan S, Ramalingam P. Biological preparation of chitosan nanoparticles and its *in vitro* antifungal efficacy against some phytopathogenic fungi. Carbohydrate Polymers. 2016;**151**:321-325. DOI: 10.1016/j.carbpol.2016.05.033

[53] Manikandan A, Sathiyabama M. Preparation of Chitosan nanoparticles and its effect on detached rice leaves infected with *Pyricularia grisea*. International Journal of Biological Macromolecules. 2016;**84**:58-61

[54] Kheiri A, MoosawiJorf SA, Malihipour A, Saremi H, Nikkhah M. Application of chitosan and chitosan nanoparticles for the control of fusarium head blight of wheat (*Fusarium graminearum*) *in vitro* and greenhouse. International Journal of Biological Macromolecules. 2016;**93**:1261-1272

[55] Khairy A, Tohamy M, Zayed M, Mahmoud S, Eltahan A, El-Saadony M, et al. Eco-friendly application of nano-chitosan for controlling potato and tomato bacterial wilt. Saudi Journal of Biological Sciences. 2021;**29**(4):2199-2209. DOI: 10.1016/j.sjbs.2021.11.041

[56] Choudhary SMK, Joshi A, Saharan V. Assessment of cu-chitosan nanoparticles for its antibacterial activity against *Pseudomonas syringae*pv. *glycinea*. International Journal of Current Microbiology and Applied Sciences. 2017;**6**(11):1335-1350

[57] Sotelo-Boyás ME, Bautista-Baños S, Correa-Pacheco ZN, Jiménez-Aparicio A, Sivakumar D. Biological activity of chitosan nanoparticles against pathogenic fungi and bacteria. In: Bautista-Banos S, Romanazzi G, Jiménez-Aparicio A, editors. Chitosan in the Preservation of Agricultural Commodities. Elsevier, USA: Academic Press; 2016. pp. 339-334

[58] Mohammadi A, Hashemi M, Hosseini SM. Chitosan nanoparticles loaded with *Cinnamomum zeylanicum* essential oil enhance the shelf life of cucumber during cold storage. Postharvest Biology and Technology. 2015;**110**:203-213

[59] Santiago T, Bonatto C, Rossato M, Lopes C, Lopes C, Mizubuti E, et al. Green synthesis of silver nanoparticles using tomato leaves extract and their entrapment in chitosan nanoparticles to control bacterial wilt: Silver and chitosan nanoparticles to control bacterial wilt. Journal of the Science of Food and Agriculture. 2019;**99**. DOI: 10.1002/jsfa.9656

[60] Li B, Zhang Y, Yang Y, Qiu W, Wang X, Liu B, et al. Synthesis, characterization, and antibacterial activity of chitosan/TiO2 nanocomposite against *Xanthomonas oryzae* pv. oryzae. Carbohydrate Polymers. 2016;**5**(152):825-831

[61] Pospieszny H, Chirkov S, Atabekov J. Induction of antiviral resistance in plants by chitosan. Plant Science. 1991;**79**:63-68

[62] Faoro F, Sant S, Iriti M, Appiano A. Chitosan-elicited resitance to plant viruses: A histochemical and cytochemical study. In: Muzzarelli RAA, editor. Chitin Enzymology. Italy: Grottammare; 2001. pp. 57-62

[63] Chirkov SN. The antiviral activity of chitosan (review). Applied Biochemistry and Microbiology. 2002;**38**:1-8

[64] Kulikov SN, Chirkov SN, Il'ina AV, Lopatin SA, Varlamov VP. Effect of the molecular weight of chitosan on its antiviral activity in plants. PrikBiokhimMikrobiol. 2006;**42**(2):224-228

[65] Lu H, Zhao X, Wang W, Yin H, Xu J, Bai X, et al. Inhibition effect on tobacco mosaic virus and regulation effect on calreticulin of oligochitosan in tobacco by induced Ca2+ influx. Carbohydrate Polymers. 2010;**82**:136-142. DOI: 10.1016/j.carbpol.2010.04.049

[66] Siddaiah CN, Prasanth KVH, Satyanarayana NR, Mudili V, Gupta VK, Kalagatur NK, et al. Chitosan nanoparticles having higher degree of acetylation induce resistance against pearl millet downy mildew through nitric oxide generation. Scientific Reports. 2018;**8**(1):2485. DOI: 10.1038/s41598-017-19016-z

[67] Kaur P, Thakur R, Choudhary A. An *in vitro* study of the antifungal activity of silver/chitosan nanoformulations against important seed borne pathogens. International Journal of Scientific and Technology Research. 2012;**1**:83-86

[68] Brunel F, El Gueddari NE, Moerschbacher BM. Complexation of copper (II) with chitosan nanogels: Toward control of microbial growth. Carbohydrate Polymers. 2013;**92**(2):1348-1356

[69] Benhamou N. Ultrastructural detection of β-1,3-glucans in tobacco root tissues infected by *Phytophthora parasitica* var. *nicotianae* using a gold-complexed tobacco β-1,3-glucanase. Physiology Molecular Plant Pathology. 1992;**41**:351-357

[70] Iriti M, Faoro F. Chitosan as a MAMP, searching for a PRR. Plant Signaling & Behavior. 2009;**4**(1):66-68

[71] Amborabé B-E, Bonmort J, Fleurat-Lessard P, Roblin G. Early events induced by chitosan on plant cells. Journal of Experimental Botany. 2008;**59**:2317-2324

[72] Zayed M, Elkafafi S, Zedan A, Dawound S. Effect of Nano chitosan of growth, physiological and biochemical properties of Phaseolus vulgaris under salt stress. Journal of Plant Production. 2017;8:577-585

Chapter 5

Chitosan Is the Ideal Resource for Plant Disease Management under Sustainable Agriculture

Magdi A.E. Abdellatef, Eman Elagamey and Said M. Kamel

Abstract

In the spirit of returning to nature and using scientific applications to raise plant efficiency and reduce pathogen risk, scientists began searching for safe, natural alternatives to pesticides that are highly effective and low cost. On top of these alternatives, chitosan came with its biodegradability, biocompatibility, antimicrobial activity, and nontoxicity, which granted it dual energetic effects during the host-pathogen interaction. Chitosan promotes plant growth, regulates plant cell homeostasis and metabolic processes, and triggers plant defense mechanisms; on the other hand, it inhibits the ability of pathogens by disrupting pathogen growth and reducing reproduction, wherefore chitosan will become an increasingly prevalent and ideal resource for agricultural sustainability.

Keywords: chitosan, eco-friendly, antimicrobial activity, abiotic stress, defense responses

1. Introduction

Diverse plant species are adversely affected by phytopathogens on a large scale, which results in severe economic loss and microbial toxins in food and animal feed. The conventional approaches to plant disease control are no longer efficient and secure, especially in light of the occurrence of severe climatic changes. Researchers had to change their traditional thinking in the process of combating plant diseases, which depends mainly on the use of chemical pesticides. In line with the strategy of sustainable agriculture, natural compounds, especially chitosan, have been put into use as safe alternatives that are effective in control, improving plant properties, and keeping the biological balance between plants and beneficial microorganisms in the soil.

Chitosan is a natural, biodegradable, and environmentally eco-friendly polysaccharide obtained from the exoskeleton of crustaceans such as shrimp, shellfish, crabs, cuttlefish, squid pen, and crawfish through the deacetylation process of chitin [1]. Moreover, chitosan can be produced from fungal chitin [2]. Chitosan has a unique nature as it is a linear polymer consisting of two subunits linked together, namely D-glucosamine and *N*-acetyl-D-glucosamine [3]. Chitosan concentration, chemical modification, acetylation degree, and molecular weight have all been identified as critical factors in the suppression of plant pathogen infection in the host plant [4, 5]. In sustainable agriculture, chitosan can be used alone or in combination with other compounds in plant nutrition with the aim of withstanding abiotic stress, stimulating plant defense systems, combating plant diseases, promoting plant seed germination and seedling growth [6, 7], and increasing crop production and quality. In addition, chitosan can purify soil and agricultural wastewater from heavy metals such as mercury, copper, uranium, and lead and thus can be reused for irrigation [8].

2. Chitosan chemical, physical, and biological features

Chitosan has many properties that have captured the interest of researchers over the past 20 years to explore the prospect of using it in a variety of scientific and practical sectors. Chitosan properties depend on several very important factors, such as molecular weight, degree of deacetylation, and solubility. The molecular weight of chitosan might influence the crystal size and morphological character of chitosan-based thin-film composites (TFCs) and other products or membranes [9]. The chitosan molecular weight may range between 50 and 2000 kDa depending on the source of chitin, and it has numerous influences on the viscoelastic properties of solutions and hydrated colloidal forms [10]. The degree of deacetylation determines the content of free $-NH_2$ groups in the polysaccharide and it has an influence on all the functional properties of chitosan [9]. Chitosan can dissolve in aqueous solutions, but it does so more readily in acidic media than in neutral or basic media [11]. The solubility of chitosan varies according to several factors, including polymer molecular weight, degree of acetylation, pH, temperature, and polymer crystallinity. Chitosan has unique chemical properties, e.g., linear polyamine, reactive amino and hydroxyl groups, and chelated metal ions. In addition, the biological properties of chitosan are biocompatibility, biodegradability, antimicrobial activity, biosafety, and nontoxicity [12]. These biological properties vary in influence on plants and their fungal and fungal-like microorganisms, bacterial, viral, viroid, and nematode pathogens according to the physical and chemical features of chitosan. Thus, chitosan compounds are strongly recommended to be used in the management strategies against phytopathogens such as viruses [4], bacteria [13], and fungi [14].

3. Antimicrobial activity of chitosan against phytopathogens

Several hypotheses have been postulated to explain the mechanism by which chitosan affects several phytopathogens:

a. Chitosan has polycationic nature enables it to interfere with electronegative charges on the outer surface of the microbial cell. The external electrostatic interaction between the positive amino glucosamine groups $-NH3^+$ of chitosan and the negative charge on the cell surface exists in teichoic acids in gram-positive bacteria, lipopolysaccharides in gram-negative bacteria, and phospholipids in the fungal cell membrane (**Figures 1** and **2**), leading to changes in cell permeability and leakage of intracellular electrolytes and proteinaceous constituents and cell death [15, 16]. Divya et al. [17] suggest that the antimicrobial properties of chitosan are due to the repeated amino groups on the backbone of the polymer structure.

b. Chitosan functions as a chelating agent of metals and vital nutrients, causing microbial starvation and impairing microbial development [18] as the amine

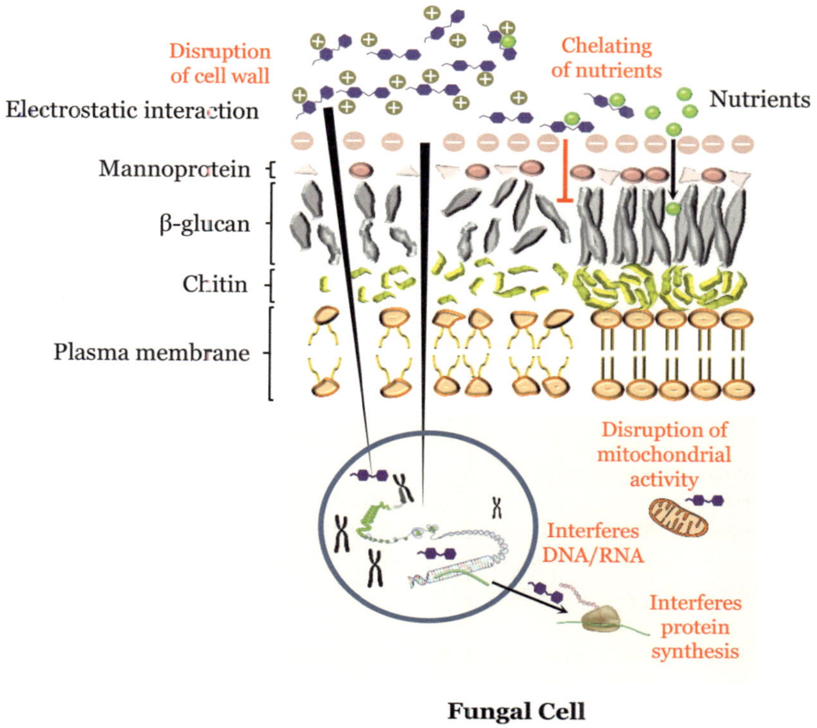

Figure 1.
Potential antifungal activity of chitosan.

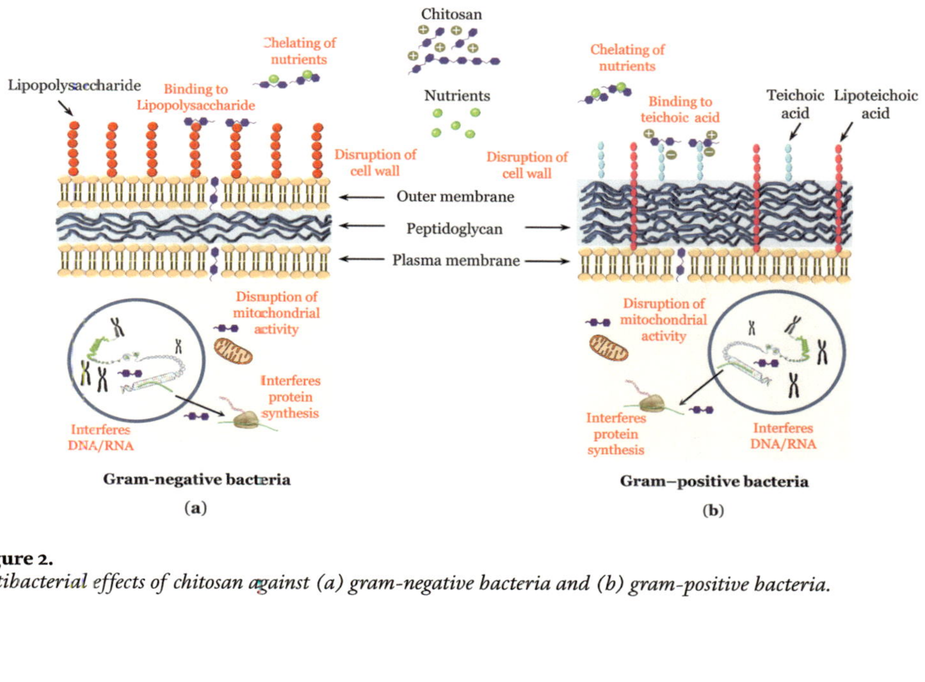

Figure 2.
Antibacterial effects of chitosan against (a) gram-negative bacteria and (b) gram-positive bacteria.

groups in the chitosan molecules are in charge of the uptake of metal cations by chelation. Chitosan metal-binding capacity increases at high pH since the amine groups are not protonated and the electron pair on nitrogen in the amine group is available for donation to metal ions [19]. The high molecular weight of chitosan might cause a reduction in cell membrane permeability due to a polymer coating on the surface of the cells that blocks cell access to nutrients [20].

c. The internal electrostatic interactions between the positive amino groups on the polysaccharide chain of low molecular weight chitosan and the negative phosphate groups on the nucleic acid chain of microbial cells lead to the inhibition of the synthesis of DNA/mRNA and a decrease in the abundance level of protein and enzymes [21].

Not all forms of pathogens are equally sensitive to chitosan (**Table 1**); therefore, the degree to which pathogens react to chitosan depends on a number of factors:

a. The solubility of chitosan increases in acidic solutions and becomes insoluble in solutions with a pH greater than 6.5, according to the pKa values of its amino groups, which range from 6.3 to 6.5. Many articles have shown that chitosan is a great antimicrobial agent under acidic conditions [21]. Meanwhile, the pathogen inhibitory effect of chitosan depends on the type of solvent. Chitosan dissolved in lactic acid shows the best inhibitory effect as compared to that dissolved in formic acid and acetic acid [22].

b. The ability of chitosan to exhibit intracellular antimicrobial action depends on its molecular weight. Low molecular weight chitosan has the highest inhibitory effect on *Rhizopus stolonifera* [23], while high molecular weight chitosan shows better efficacy on *Fusarium oxysporum* f. sp. *vasinfectum*, and *Alternaria solani* [25].

   Furthermore, the abundance of polysaccharides and proteins that make up the numerous layers of the cell wall in both fungi and bacteria has a major impact on the mechanical strength of interaction with chitosan [73].

c. The degree of deacetylation affects the antimicrobial properties of chitosan. Chitosan with a high degree of deacetylation exhibits a more positive charge, which results in stronger electrostatic interactions with the microbial cell surface and higher antimicrobial activity [13].

d. The inhibition rate of pathogen growth depends on the concentration of chitosan [74].

3.1 Antifungal activity

Chitosan is efficient in inhibiting hyphal growth, mycelial elongation, spore formation, spore germination, spore viability, germinal tube, and fungal virulence factor production of phytopathogenic fungi [14]. The ability of chitosan to penetrate the plasma membranes of phytopathogens depends on the degree of membrane fluidity. Chitosan-sensitive fungi possess polyunsaturated fatty acid-rich membranes such as linoleic acid (high fluidity membrane), while chitosan-resistant fungi possess saturated

Host/disease name	Pathogen/ stress	Chitosan effects	Chitosan mode of action	References
Tomato wilt	*Fusarium oxysporum* f. sp. *radicis-lycopersici*	Antifungal	Chitosan was bound to negatively charged phospholipids that alter plasma membrane fluidity and induced membrane permeabilization.	[15]
Potato dry rot	*Fusarium sulphureum*	Antifungal	Chitosan caused morphological changes such as intertwisting hyphal, distortion, and swelling with excessive branching, abnormal distribution of cytoplasm.	[22]
Post-harvest fungi of fruits and vegetables	*Rhizopus stolonifer*	Antifungal	Low molecular weight chitosan caused inhibition of mycelial growth, while the high molecular weight chitosan affected spore shape, sporulation, and germination.	[23, 24]
Early blight of potato, tomato, capsicum, and eggplant	*Alternaria solani*	Antifungal	Chitosan reduced hyphal growth, inhibited sporulation and spore germination, and induced morphological changes.	[25, 26]
Cotton wilt	*Fusarium oxysporium* f. sp. *vasinfectum*	Antifungal		
Apple dieback	*Valsa mali*	Antifungal		
Banana wilt	*Fusarium oxysporum* f. sp. *cubense*	Antifungal	Chitosan formed abnormal shapes, vesicles, and empty cells devoid of cytoplasm in the mycelia and agglomeration of hyphae.	[27]
Soybean sudden death syndrome	*Fusarium solani* f. sp. *glycines*	Antifungal	Chitosan was able to induce the level of chitinase activity in soybean. Furthermore, chitosan interfered directly with the fungal membrane function and interacted with fungal DNA and mRNA.	[28]
Fruit rots in mango, strawberry, apple, and peach	*Alternaria alternata*	Antifungal	Chitosan caused aggregation of mycelium and structural changes such as excessive branching, cell wall swelling, and hyphae size reduction.	[24, 29]
	Botrytis cinerea	Antifungal		[24]
	Penicillium expansum	Antifungal		[24]
Bayoud disease in date palm	*Fusarium oxysporum* f. sp. *albedinis*	Antifungal	Chitosan elicited a defense reaction against this fungus in date palm roots, by eliciting peroxidase and polyphenoloxidase activity and increasing the level of phenolic compounds.	[30]

Host/disease name	Pathogen/ stress	Chitosan effects	Chitosan mode of action	References
Cucumber wilt	*Fusarium oxysporum* f. sp. *cucumerinum*	Antifungal	Chitosan caused hindering of the Fusarium cell wall, disrupting DNA, disrupting structural and functional protein biosynthesis, and influencing metabolic pathways.	[31]
Early blight disease	*Alternaria solani*	Antifungal	Chitosan covered the cell wall, causing membrane disruption and cell leakage. Moreover, chitosan penetrated into fungal living cells, leading to the inhibition of various enzymes and interference with the synthesis of mRNA and proteins.	[29]
Root rot disease	*Fusarium oxysporum*	Antifungal		
Damping off disease	*Pythium debaryanum*	Antifungal		
Postharvest decay of oranges	*Penicillium citrinum*, *Penicillium mallochii*	Antifungal	Chitosan inhibited the growth of two *Penicillium* species by compacting fungal spores and reducing the activity of plant cell walls by degrading enzymes produced by fungi. Additionally, chitosan increased the wound healing process in orange fruits.	[32]
Anthracnose and ripen fruit rot in chilli	*Colletotrichum capsici*	Antifungal	Chitosan completely inhibited the mycelial growth of *C. capsici* by forming physical barriers around the penetration sites of the pathogen, preventing them from spreading to healthy tissues. In addition, chitosan activated the defense-related antimicrobial compounds, induced a decline in malondialdehyde content, and increased the concentrations of soluble sugars and proline, as well as peroxidase and catalase activities.	[33]
Anthracnose of camellia	*Colletotrichum camelliae*	Antifungal	Chitosan enhanced the activity of H_2O_2, the defense-related enzymes, such as polyphenol oxidase, peroxidase, catalase, phenylalanine ammonia-lyase, and the concentration of soluble proteins inside the camelia plant.	[34]

Host/disease name	Pathogen/ stress	Chitosan effects	Chitosan mode of action	References
Fasciation of *Arabidopsis thaliana*	*Staphylococcus simulans*	Antibacterial	Chitosan caused multiple changes in the expression profiles of *Staphylococcus aureus* SG511 genes involved in the regulation of stress and autolysis, as well as genes associated with energy metabolism were observed.	[35]
Common scab in potato tubers	*Streptomyces scabies*	Antibacterial	Chitosan reduced disease incidence and disease severity by eliciting plant defense mechanisms and systemic resistance.	[36]
Bacterial leaf blight and leaf streak in rice	*Xanthomonas oryzae* pv. *oryzae, Xanthomonas oryzae* pv. *oryzicola*	Antibacterial	Chitosan caused membrane lysis and the destruction of biofilm of bacteria and increased the activities of phenylalanine ammonia-lyase, peroxidase, and polyphenol oxidase in rice seedlings.	[37]
Bacterial leaf spot of tomato	*Xanthomonas vesicatoria*	Antibacterial	Chitosan induced systemic resistance mechanisms in tomato by activation of several defense enzymes and elicited expression of PIN II and ETR-1 genes from several molecular pathways involved in pathogen defense.	[26]
Bacterial head rot of broccoli	*Pseudomonas fluorescens*	Antibacterial	Chitosan reduced the disease incidence, the lesion diameter and induced systemic resistance mechanisms in broccoli.	[38]
Bacterial brown stripe of rice	*Acidovorax avenae* subsp. *avenae*	Antibacterial	Chitosan caused membrane disruption and lysis, reduction of biofilm formation, and gene expression change.	[39]
Bacterial fruit blotch of watermelon	*Acidovorax citrulli*	Antibacterial	Chitosan significantly inhibited the growth of *A. citrulli* by damaging and altering the cell membrane, separating the cytoplasmic membrane from the cell envelope, coagulating the cytosolic components, forming a vacuole-like structure, and broking of cell walls, leading to the leaching out of a mass of nutrient and nucleic materials.	[40]

Host/disease name	Pathogen/ stress	Chitosan effects	Chitosan mode of action	References
Bacteria of crown gall disease	*Agrobacterium tumefaciens*	Antibacterial	Chitosan was bound with negatively charged components on the bacterial surface, via electrostatic interactions, causing changes in the permeability of the bacterial wall and shutting down the cell division, leading to bacterial fatality.	[29]
Soft mold disease	*Erwinia carotovora*	Antibacterial		
Bean mild mosaic virus	BMMV	Antiviral	Chitosan led to the inhibition of systemic propagation of viruses over bean plants and enhanced the hypersensitivity response of bean plants to viral infection.	[5]
Tobacco mosaic virus in *Arabidopsis thaliana*	TMV	Antiviral	Chitosan induced TMV resistance in wild-type and jasmonic acid pathway-deficient (jar1) Arabidopsis plants enhanced the expression of the defence-related gene PR1.	[41]
Alfalfa Mosaic Virus in *Nicotiana glutinosa*	AMV	Antiviral	Chitosan induced systemic acquired resistance, increased total carbohydrates and total phenolic contents, as well as triggered the transcriptional levels of peroxidase, pathogen-related protein-1, and phenylalanine ammonia-lyase.	[42]
Banana root-knot nematode	*Meloidogyne incognita*	Antinematode	Chitosan suppressed root-knot nematodes by decreasing the total number of galls, immature stages, females with egg masses, and the count and rate of build-up of nematodes in the soil.	[43]
Tomato root-knot nematode	*Meloidogyne javanica*	Antinematode	Chitosan reduced the number of nematodes per plant and promoted plant growth.	[44]
Carrot root-knot nematode	*Meloidogyne incognita*	Antinematode	Chitosan caused a synergistic effect against *M. incognita* on carrots and could be utilized to manage nematodes as a potential sustainable treatment.	[45]
Rice root-knot	*Meloidogyne graminicola*	Antinematode	Chitosan reduced root-galling and nematodes development and induced a plant defense mechanism against nematodes.	[46]

Host/disease name	Pathogen/ stress	Chitosan effects	Chitosan mode of action	References
Thymus daenensis	Abiotic stress	Drought stress	Chitosan stimulated osmotic adjustment through proline accumulation and reduction of lipid peroxidase levels, which increased the integrity of cell membranes of thyme leaves.	[47]
Potato plants	Abiotic stress	Drought stress	Chitosan reduced membrane relative permeability and malondialdehyde concentration of potato leaves; raised the concentration of proline and soluble proteins; enhanced the activities of superoxide dismutase, and enhanced their antioxidation ability; increased the activities of protective enzymes and regulated the content of osmotic regulatory substances like peroxidase.	[48]
Apple seedlings	Abiotic stress	Drought stress	Chitosan acted as an exogenous antioxidant that enhanced resistance to oxidative stress during drought in apple seedlings.	[49]
Cowpea plants	Abiotic stress	Drought stress	Chitosan increased the thickness of the midrib region, mesophyll tissue, and the midrib vascular bundle, therefore reduced water stress in cowpea.	[50]
Sweet basil (*Ocimum ciliatum*, *Ocimum. basilicum*)	Abiotic stress	Drought stress	Chitosan reduced the harmful effect of water stress, increased plant growth parameters, and significant effected on total phenol content and antioxidant activity of the extracts of the two species.	[51]
Coffee plants	Abiotic stress	Drought stress	Chitosan enhanced the content of chlorophyll and carotenoids in the leaves of coffee seedlings, increased mineral uptake and stimulated the growth of coffee seedlings.	[52]
Pepper plants	Abiotic stress	Antitranspirant	Chitosan decreased transpiration due to inducing closure of the plant's stomata by a decrease of K in the guard cells.	[53]
Safflower and Sunflower seedlings	Abiotic stress	Anti-salinity	Chitosan exhibited positive effects on salt stress alleviation through the reduction of MDA and increased proline contents and CAT and POX activities in both crops.	[54]

Host/disease name	Pathogen/ stress	Chitosan effects	Chitosan mode of action	References
Wheat seedlings	Abiotic stress	Anti-salinity	Chitosan increased antioxidant enzyme: superoxide dismutase (SOD), catalase (CAT) and peroxidase (POX) activities, reduced malondialdehyde (MDA) content in leaves, and accelerated the accumulation of proline.	[55]
Maize plants	Abiotic stress	Anti-salinity	Chitosan ameliorated the effects of salinity stress and improved plant growth.	[56]
Rice seedlings	Abiotic stress	Anti-salinity	Chitosan lowered malondialdehyde and increased proline levels and enhanced the activities of catalase and peroxidase enzymes.	[57]
Fenugreek (Trigonella foenum-graecum)	Abiotic stress	Anti-salinity	Chitosan increased plant dry weight, length of stem and roots, leaf relative water content, and amount of anthocyanin photosynthetic pigments, resulting in better growth and establishment of fenugreek plants.	[58]
Bean (*Phaseolus vulgaris*)	Abiotic stress	Anti-heat stress	Chitosan enhanced uptake of mineral nutrients, cell division, and chlorophyll accumulation in the leaves.	[59]
Cucumber seedlings	Abiotic stress	Anti-heat stress	Chitosan improved the cold resistance of cucumber seedlings, protected the membrane system, improved the capability of eliminating active oxygen species, and alleviated the damage to photosynthetic organization.	[60]
Maize plants	Abiotic stress	Anti-heat stress	Chitosan caused a decline in malondialdehyde (MDA) content and relative permeability of the plasma membrane. Also, increased concentrations of soluble sugars, proline, and peroxidase (POD) and catalase (CAT) activities led to an enhanced germination index and reduced the mean germination time (MGT). Additionally, it increased shoot height, root length, and shoot and root dry weights.	[61]

Host/disease name	Pathogen/ stress	Chitosan effects	Chitosan mode of action	References
Bean (*Phaseolus vulgaris*)	Plant improvement	Promote plant growth	Chitosan enhanced shoot and root length, fresh and dry weights of shoots and roots, and leaf area in beans.	[62]
Rice and soybean plants	Plant improvement	Promote plant growth	Chitosan promoted plant length and shoot growth in both rice and soybean.	[63]
Cucumber plants	Plant improvement	Promote plant growth	Chitosan increased the vegetative growth and increased the total yield of cucumber fruits in terms of quantity and quality.	[64]
Okra plants	Plant improvement	Promote plant growth	Chitosan enhanced the biochemical activities in okra, resulting in an increase in growth and the fruit-bearing nodes, and improved the yield quality and quantity.	[65]
Tomato and eggplant plants	Plant improvement	Promote plant growth	Chitosan enhanced the growth and functional components of tomatoes and eggplants by increasing plant height, branch, leaf, flower, and fruit number and size.	[66]
Verbena bonariensis	Plant improvement	Promote plant growth	Chitosan enhanced the plant height, number of inflorescences and number of leaves.	[67]
Eustoma grandflorum	Plant improvement	Promote plant growth	Chitosan soil treatment resulted in greater shoot length, stem diameter, weight of cut flowers, and an increase in the number of flowers.	[68]
Soybean leaves	Plant improvement	Stimulate plant defense responses	Chitosan increased activity of phenylalanine ammonia-lyase and tyrosine ammonia-lyase enzymes which increased total phenolic content in plants.	[69]
Chickpea plants	Plant improvement	Imparts immunity against Fusarium	Chitosan caused significant ECM and guard cell remodelling, and translated ECM cues into cell fate decisions during fusariosis.	[70]
Several plants	Plant improvement	Eliciting plant immunity	Chitosan-induced defense responses such as increased cytosolic H^{+} and Ca^{2+}, MAP-kinase activation, callose apposition, oxidative burst, and hypersensitive response (HR) Moreover, it induced the synthesis of abscisic acid, jasmonate, phytoalexins, and pathogenesis-related (PR) proteins.	[71]

Host/disease name	Pathogen/ stress	Chitosan effects	Chitosan mode of action	References
Soybean cells	Plant improvement	Elicited callose synthesis	Chitosan caused the rapid and quantitative formation of callose and elicited other metabolic changes that appeared to be important for resistance against pathogens in soybean cells.	[72]

Table 1.
Chitosan mechanisms in plant protection against biotic and abiotic stresses.

fatty acid-rich membranes such as palmitic acid or stearic acid (low fluidity membrane) [15]. The incubation period affects the antifungal activities of chitosan. The growth inhibition of *F. oxysporum* f. sp. *radicis-lycopersici* at a low concentration of chitosan increases with the long incubation period [74]. Chitosan causes excessive mycelial branching, abnormal shapes, swelling and hyphae size reduction in *F. oxysporum* f. sp. *cubese* [27], *F. solani* f. sp. *glycines* [28], *Botrytis cinerea*, and *A. alternate* [24]. Chitosan is also responsible for cytological alteration, protoplasm dissolution, and large vesicles of fungus [27]. Chitosan caused morphological changes such as large vesicles or empty cells devoid of cytoplasm in the mycelium of *B. cinerea* and *F. oxysporum* f. sp. *albedinis* [30, 75]. Fungi that have been exposed to chitosan make fewer spores than untreated fungi. In other instances, full sporulation suppression was observed following chitosan therapy. The length and shape of the conidia of *Ranunculus stolonifer*, *Penicillium digitatum*, and *F. oxysporum* were considerably influenced by chitosan [76]. The application of chitosan against *F. oxysporum* f. sp. *cucumerinum* showed a significant decrease in wilt disease severity compared to chitosan-untreated Fusarium [31].

3.2 Antibacterial activity

Gram-positive and gram-negative bacteria have significantly different cell wall structures and surface polarity, which results in different sensitivity to chitosan [13]. The cell wall of gram-negative bacteria is distinguished by the presence of lipopolysaccharides, which contain phosphate and pyrophosphate groups [77]. This provides it with a high negative charge, making it more bound to chitosan [78]. While the cell wall of gram-positive bacteria contains polysaccharides associated with lipoteichoic and teichoic acids. Teichoic acid is negatively charged due to the presence of phosphate groups in its structure which gives it a small negative charge that makes it less bound to chitosan. The stopping of the teichoic acid biosynthesis pathway in *Staphylococcus aureus*, led to an increase in chitosan resistance [35]. Gram-negative bacteria are more hydrophilic than gram-positive bacteria and also have a thinner cell membrane [79]. This explains why gram-negative bacteria react to chitosan differently from gram-positive bacteria (**Figure 2**).

Chitosan has potent antibacterial activities against a variety of plant pathogenic bacteria like *S. aureus* [80], *Streptomyces scabies* [36] *Ralstonia solanacearum* [81], *Xanthomonas* spp. [26, 37], *Pseudomonas* spp. [38, 80], and *Acidovorax* spp. [39, 40]. The inhibitory activity of chitosan against bacteria varied with molecular weight [82], concentration [39], solvent type [29], bacterial type (gram-positive/gram-negative) [83], cell wall structure [84], period of incubation and abiotic factors [85]. Chitosan binds to the negatively charged surface of bacteria at low concentrations (less than

0.2 mg/ml) to cause agglutination, but at higher concentrations, the presence of more chitosan positive charges may have given the bacteria a net positive charge that keeps them suspended [86]. Moreover, Goy et al. [87] suggest that chitosan is responsible for the hydrolysis of peptidoglycans, the main component of the bacterial cell wall, increasing electrolyte leakage, and potentially causing the death of the plant pathogens. Additionally, Liang et al. [88] reported that chitosan is the responsible substance for the destruction of the bacterial cell membrane, which causes death due to the leakage of intracellular substances. Chitosan applied to tomato plants inhibited the growth of *Xanthomonas vesicatoria* [26]. Chitosan-protected cucumber from *Pseudomonas syringae* pv. *lachrymans* that causes bacterial angular leaf spot damage [89]. Chitosan has decreased the disease incidence of broccoli that was infected with *Pseudomonas fluorescens* [38]. The disease index of watermelon seedlings infected with *Acidovorax citrulli* was significantly reduced by chitosan at 0.4 mg/ml [40]. Chitosan solution at 0.10 mg/ml markedly decreased the surviving cell number of *Xanthomonas* pathogenic bacteria isolated from different geographical origins compared with the control after 6 h of incubation, regardless of the bacterial strain [90].

3.3 Antiviral activity

One of the most destructive plant diseases is the viral disease which causes serious damage to many plant species, affecting agroecosystems and food security. For that reason, searching for new eco-friendly application technologies to suppress the invasion of viral plant diseases is urgently needed to fulfill the nutrients required to feed the world's population [91]. Chitosan and its derivatives have been used as a promising and powerful tool against plant viruses. Chitosan has demonstrated antiviral activity against Potato virus X (PVX) through the possible mechanism of induced resistance and responsive defense mechanisms or the inhibition of systemic propagation of plant viruses in potato plants [4]. Complete inhibition or suppression of systemic virus multiplication in the host plant has not clearly proven the capability of chitosan to stop the virus activation. However, the multiplication block may be due to the binding of chitosan molecules with the nucleic acid of a targeted virus, causing serious damage to the viral genome [92]. Studies by Jia et al. [41] explained the role of chitosan in inducing systemic acquired resistance in Arabidopsis plants infected with Tobacco mosaic virus (TMV) and which signaling pathways are involved in the processes of defense mechanisms. Their obtained results revealed that chitosan application induced TMV resistance through specific pathways in the plant. The induction in the Arabidopsis plants happened through the jasmonic acid pathway-deficient (Arabidopsis plants jar1) and at the same time did not induce the salicylic acid pathway-deficient (Arabidopsis plants NahG). The application of chitosan as a protective and curative treatment against Alfalfa mosaic virus (AMV) on *Nicotiana glutinosa* plants under greenhouse conditions was studied by [42]. They proved that the AMV concentration was significantly reduced through both protective and curative treatment with 70.43% and 61.65%, respectively. On the other hand, possible ways of inducing systemic resistance and responsive defense mechanisms were measured, resulting in an increase of total phenols and carbohydrates as well as phenylalanine amonialyase (PAL) and peroxidase.

3.4 Antinematode activity

Serious, highly damaging, and economic losses to a wide range of plant hosts, including fruit trees, vegetables, agronomic crops, foliage crops, grasses, nuts, and forest trees, were reported to be caused by several nematode species under different genera. Economic losses to more than 2000 kinds of higher plant species are often great when certain plant species are grown in warm regions around the world. The development of efficient new and eco-friendly nematode management strategies such as biological control, natural products, plant extracts, and botanical products is needed urgently to reduce the high toxicity of chemical nematicides [93]. Studies by El-Ansary et al. [43] revealed that chitosan significantly reduced the disease severity of root-knot nematode infection caused by *Meloidogyne incognita* in banana plants cv. Williams with an improvement in plant growth parameters and yield production. Chitosan-treated tomato plants produced less root-knot nematode reproduction, which enhanced the size, weight, and growth of the plant's roots and shoots [44]. Recently, Khan et al. [45] evaluated chitosan as a nematicide against the infestation caused by *Megalaima incognita* in carrot plants under *in vivo* and *in vitro* conditions. They reported that egg masses and second-stage juvenile (J2s) of *M. incognita* were affected by the usage of different concentrations of 500, 1000, 1500, 2000, and 2500 ppm of chitosan. Maximum mortality of J2s and the highest inhibition in egg hatching was observed at 2500 ppm of chitosan after 36 h incubation period.

4. Role of chitosan in plant protection against abiotic stress

Chitosan improves plant tolerance to drought, salinity, and high temperature [94] (**Table 1**).

4.1 Effect on drought stress

Agricultural productivity is limited by drought stress, which has a number of negative effects on plant health and lowers plant growth and yield. Chitosan is one of the effective solutions to mitigate the harmful effects of drought stress on plants (**Figure 3a**). Under extreme drought stress, the free proline content in leaves considerably increases [95]. In chitosan-treated plants, the accumulation of proline production in the absence of drought stress was enhanced [96]. After chitosan treatment, proline accumulation was enhanced in the thyme plants [47]. The accumulation of proline helps in reducing the leaf water potential, improves the turgor of leaves, and facilitates water delivery to them. In addition, proline is crucial for maintaining redox balance, quenching ROS, and osmotic adjustment [97].

On the other hand, a water deficit condition disturbs the plant's cell membrane integrity. The membrane permeability and malondialdehyde concentration are related to each other, indicating membrane stability. Malondialdehyde levels rise in conditions of water deficiency; this lipid peroxidation byproduct has the potential to lead to membrane leakage because of the accumulation of free radicals. Chitosan functions as a positive regulator in osmotic adjustment and can eliminate the adverse effects of drought stress symptoms by decreasing the production of lipid peroxidation. The pretreatment of thyme and potatoes with chitosan reduced lipid peroxidation, removed ROS, and improved cell membrane integrity [47, 48]. In apple seedling leaves exposed to drought stress, chitosan significantly improved the integrity of cell membranes and decreased the production of malondialdehyde and electrolyte leakage [49]. Plants attempt to reduce the harmful effect of drought by raising the level of soluble sugar in the cell by breaking down the polysaccharides that help in the preservation of turgor [98]. Chitosan is an important source of sugars, e.g., glucose, fructose, trehalose,

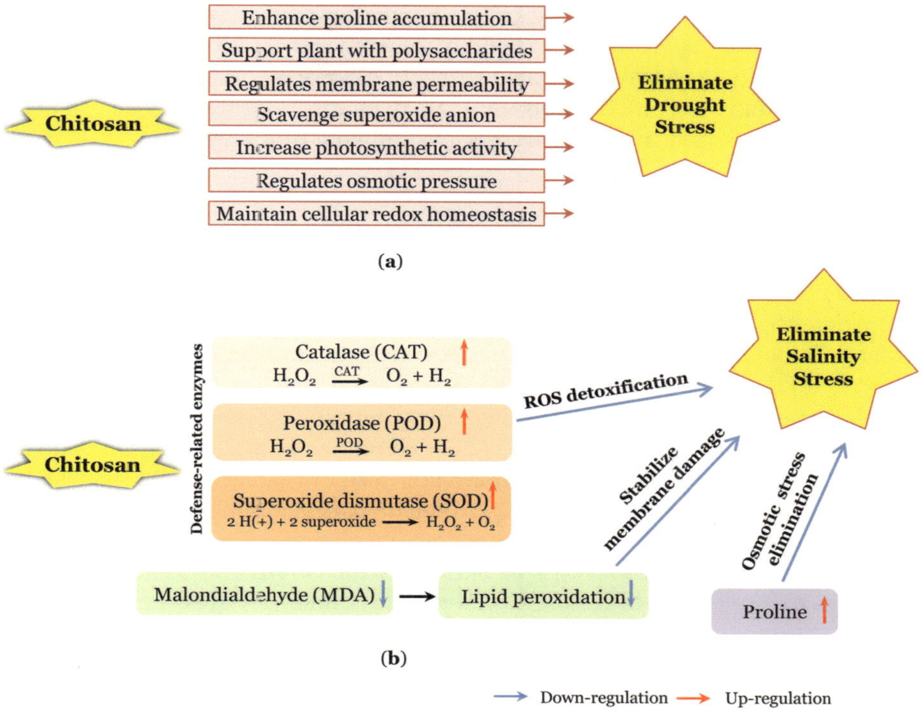

Figure 3.
Role of chitosan in removing (a) drought stress and (b) salinity stress.

sorbitol, mannose, and myoinositol that plants need to overcome drought [96]. These might enhance osmotic adjustment and maintain carbon balance in response to dehydration stress, which would improve drought resistance.

Additionally, drought stress hinders photosynthetic activity by reducing chlorophyll production. This could be caused by oxidative damage to chloroplast lipids, pigments, and proteins or by the loss of chlorophyll pigment complexes or light-harvesting protein complexes [50, 99]. Chitosan spraying resulted in an increase in chlorophyll and total carbohydrates which increased photosynthesis levels in soybean and maize [100], cowpea [50], and bean [62]. This might be the result of higher nitrogen and potassium levels in plant shoots, which aid in raising the number of chloroplasts per cell and boosting chlorophyll synthesis. Additionally, chitosan treatment's release of amino compounds with a higher availability level encourages the synthesis of chlorophyll [63].

Commercial antiperspirants are better than chitosan in raising the efficiency of the plant in retaining water, but they reduce the photosynthetic rates and carbon uptake as a result of reducing the internal CO_2 in leaves [71]. The efficiency of chitosan as an antiperspirant is due to its control of the mechanism of opening and closing the stomata, which allows the entry of carbon from the atmosphere into the plant. Thus, maintaining the efficiency of the photosynthesis process inside the plant, unlike the commercial antitranspirant, this acts as a thin antitranspirant membrane that covers the leaves' surface and blocks stomata, which prevents the entry of carbon needed for photosynthesis. Chitosan use as an antitranspirant substance would be more suitable for plants that experience occasional drought occurrences. Chitosan-treated plants would enable their natural physiological system to quickly recover maximum carbon uptake while sustaining biomass and yield in these circumstances [53]. Histological examination of chickpea leaves revealed stomatal closure accompanied by decreases in stomatal conductance and transpiration rate in chitosan-treated seedlings during Fusarium infection, indicating the presence of stomatal immunity associated with chitosan [70].

4.2 Effect on salinity stress

Salinity can prevent plants from absorbing water and nutrients due to low external osmotic potential. In addition, the direct ionic effect results in excessive accumulation of Na and Cl ions, which causes toxic effects, closes the stomata, lowers internal CO_2, and decreases the rate of photosynthesis. Salt-induced stress conditions were discovered to have lipid peroxidation brought on by the buildup of malondialdehyde [101]. A decrease in malondialdehyde content after chitosan treatment stabilized membrane damage and may have given salt stress tolerance [54]. Plants have developed their own inherent ROS scavenging mechanism by producing antioxidant enzymatic compounds, e.g., superoxide dismutase, peroxidase, and catalase; increased abundance of these enzymes denotes effective ROS detoxification (**Figure 3b**). These enzymes were shown to be elevated in chitosan-treated plants, and they are crucial for reducing salt stress since they are stronger antioxidant enzymes [54, 55]. Chitosan has the ability to scavenge superoxide anions due to the presence of hydroxyl and amino groups that react with ROS [102]. Chitosan treatment reduces malondialdehyde levels and increases antioxidant enzyme activities during salinity stress, which minimizes the negative effects of salt stress on maize [56]. The low concentration of chitosan treatment could counteract the harmful effects of salt stress. A small amount of chitosan applied to sunflower seeds can suppress enzyme activity and decrease the oxidative damage brought on by salt stress [54]. During salt stress, wheat seed treated with chitosan showed increased levels of the antioxidant enzymes (superoxide dismutase, peroxidase, and catalase), stomatal conductance, and photosynthetic rate [55]. When a plant is exposed to salt stress, its chlorophyll concentration drastically decreases because of the accumulation of chlorophyllase and the instability of protein complexes. In salt stress, proline levels were elevated, which might be the result of increased proline biosynthesis, decreased proline oxidation to glutamate, or decreased utilization of synthesized proline. Plants are mostly protected against osmotic stress by proline [103]. Proline levels were shown to increase as chitosan concentrations increased [54].

4.3 Effect on heat stress

There is a paucity of published data on the use of chitosan in heat stress. Heat stress is frequently viewed as a complex issue because it frequently occurs in conjunction with drought stress and is challenging to measure [57]. Foliar application of chitosan combined with humic acid and zinc is the best treatment that could be recommended for dry bean production to withstand heat stress [59]. Chitosan use could reduce the effects of high temperatures by promoting abscisic acid activity [53]. According to Choi et al. [104], abscisic acid can activate genes associated with heat shock, such as ABF3, whose overexpression may improve heat stress tolerance. Applying chitosan to cucumber leaves at low temperatures can improve their proline and soluble protein levels as well as the activity of antioxidant enzymes [60].

5. The role of chitosan in enhancing plant traits

When a pathogen attacks a plant, a coordinated signaling mechanism is induced which leads to the accumulation of several gene products. Once the pathogen gets recognized by plant receptors, rapid localized cell death will be developed, which is known as the hypersensitive response, causing necrosis at the site of infection. While in the sections of the plants that are not infected, a systemic expression of a broad spectrum of resistance will be induced to prevent additional pathogen infection. Then reactive oxygen species are produced, defense-related genes are activated, and the expression of genes that produce compounds, including terpenes, phytoalexins, defense enzymes, and pathogenesis-related proteins, is increased.

Chitosan can induce plant resistance and activate several defense processes in plant tissue [32]. These defense mechanisms include accumulation of hydrolytic enzymes, synthesis of proteinase inhibitors and pathogenesis-related (PR) proteins, enhancement of phytoalexins, formation of callose, promotion of lignification, and induction of reactive oxygen species (ROS) (**Figure 4**) [46, 105].

5.1 Plant growth promotion

Chitosan promotes plant growth in a variety of crops by significantly influencing the development rates of shoots, roots, flowering, and the number of flowers. Due to the great hydrophilicity of chitosan molecules, it adjusts the osmotic pressure in plant cells by increasing the absorption of water and important nutrients [61] and minimizing stress damage in plant cells. Chitosan increases the efficiency of plant nitrogen uptake, acts as a source of energy and an additional carbon source for the synthesis of carbohydrates, and acts as an activator for various metabolic processes [14, 51]. Chitosan encourages the proliferation of root cells by activating auxin and cytokinin, which further boosts nutrient uptake [52]. Chitosan assisted in triggering the hydrolytic enzymes required for the mobilization and degradation of reserve food components, including protein and starch [106]. Significant growth enhancements have been recorded after chitosan application by several studies in artichoke [107], cucumber [64], okra, [65], eggplant and tomato [66], strawberry [108], potato [109], chili [33], and watermelon [110]. Chitosan can enhance plant physiological mechanisms, e.g., nutrient absorption, cell elongation, cell division, enzymatic activation, and protein synthesis [111].

5.2 Increase photosynthetic activity

Chitosan improves the photosynthesis process by enhancing stomatal function and reducing the breakdown of chlorophyll [112]. Chitosan increased the chlorophyll levels in leaves by 13.4% compared to control plants [67]. The application of chitosan improved chlorophyll content and plant productivity in chili [113], peanut, and coffee plants [52]. The use of chitosan protects chlorophyll content in stressful circumstances; it increases the chlorophyll content in fenugreek under a salinity conditions [58]. Applying chitosan to cucumber leaves at low temperatures can improve their proline and soluble protein levels as well as the activity of antioxidant enzymes [60].

5.3 Up-regulate pathogenesis-related proteins

Plants produce proteins known as pathogenesis-related proteins (PR) in response to a pathogen invasion. They are induced as a part of systemic acquired resistance

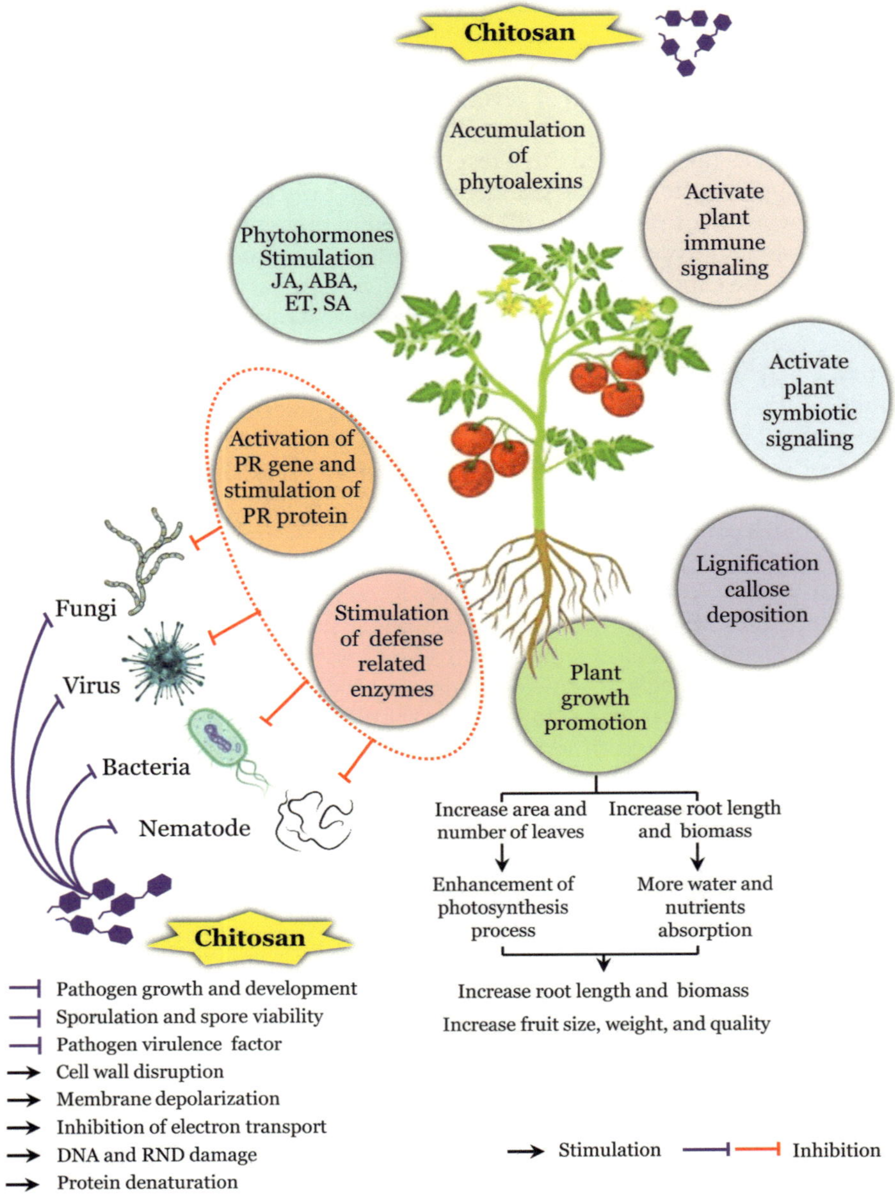

Figure 4.
Mechanisms of chitosan in the enhancement of plant defense system and inhibition of phytopathogens.

where their corresponding genes are activated by infections. Chitosan has been characterized as an elicitor that causes plants to create a wide variety of pathogenesis-related proteins with antimicrobial action. Among these pathogenesis-related proteins are chitinase and 1,3-glucanase, two hydrolytic enzymes that destroy pathogen cell walls that contain chitin and/or β-D-glucans as major structural components [114].

Chitosan appeared to use a variety of mechanisms to enhance pathogenesis-related gene function, including activating membrane receptors and altering the DNA structure of the plant. Chitosan of low molecular weight was more effective at inducing the defense-related genes β-1,3-glucanase and chitinase than the higher molecular weight [115]. Chitosan triggered the transcriptional up-regulation of defense-related genes β-1,3-glucanase and chitinase in rice seedlings [115]. Chitosan significantly increased the expression of general defense response genes in oat leaves [116]. Chitosan was able to promote resistance in pears and peaches by increasing chitinase and β-1,3-glucanase activities [74, 117]. Chitosan has been proven to be effective at triggering plant defense mechanisms by increasing the level of β-1,3 glucanase and chitinase enzymes in strawberry and pepper plants [118].

5.4 Stimulate defense-related enzymes

Chitosan acts as a physiologic elicitor, enhancing defense-related enzymes, e.g., peroxidase, catalase, superoxide dismutase, polyphenol oxidase, and phenylalanine ammonia-lyase [119]. Peroxidase helps oxidize phenolic and enodiolic compounds into quinones and hydrogen peroxide [120]. Peroxidase increases pathogen resistance in plants [121]. The chitosan treatment markedly boosted the peroxidase activity in the flesh surrounding the pear fruit wound [74]. Chitosan induced peroxidase expression activity in date palm roots when injected at three concentrations (0.1, 0.5, and 1 mg/ml) [30]. Fruit treated with chitosan maintained relatively higher peroxidase gene expression than control fruit [117]. Catalase is crucial for plant senescence and defense [122], it transforms H_2O_2 into H_2O and O_2. Catalase activity was increased in the chilling-sensitive and chilling-tolerant maize seedlings after chitosan treatment [61]. Superoxide dismutase destroys radicals and protects cells against the effects of oxidative stress. Superoxide dismutase activity has increased after chitosan treatment of *Hydrilla verticillata* [123]. Drought stress decreased the activities of the antioxidant enzymes catalase and superoxide dismutase in apple leaf tissues, but chitosan treatment enhanced their activities [49].

Polyphenol oxidase participates in plant defense by encouraging the production of lignin, which strengthens the cell wall structure and deters disease penetration [34]. Chitosan has significant antibacterial efficacy against rice leaf streak and leaf blight produced by *X. oryzae* pv. *oryzicola* and *X. oryzae* pv. *oryzae*, respectively, by increasing polyphenol oxidase activity [37]. Phenylalanine ammonia-lyase transforms L-phenylalanine to ammonia and trans-cinnamic acid [124]. It is induced in host tissues as a result of pathogen infection [69]. In grape berries, rice, and wheat, elicitation with chitosan resulted in an increase in phenylalanine ammonia-lyase [37]. When chitosan was injected into the roots of date palm, it enhanced the essential components of the host resistance against *F. oxysporum* f. sp. *albedinis*, increased the level of phenolic compounds, and stimulated date palm peroxidase and polyphenol oxidase activities [30].

5.5 Induce signal regulation

5.5.1 Activation of signal transduction

Chitosan can trigger the plant's defense mechanisms and functions as a regulatory molecule in signal transduction through several signaling pathways. When chitosan activates a particular receptor located on the cell membrane or intracellular, one or more second messengers relays the signal to the cell. This triggers a range of physiological responses as a single signal can be amplified and develop a complex signaling network. This is called signal transduction. The chitosan-mediated signal pathway includes reactive oxygen species (ROS), Ca^{2+}, nitric oxide (NO), salicylic acid (SA), abscisic acid (ABA), jasmonic acid (JA), and ethylene (ET) [125].

One of the first reactions to a microbial pathogen attack is the oxidative burst, which has been demonstrated to occur upon chitosan elicitation. It is characterized by the quick and temporary creation of enormous levels of ROS (hydrogen peroxide (H_2O_2), hydroxyl radicals (OH^-), singlet oxygen ($^1O_2$), and superoxide (O^{-2}) (**Figure 5**) [126]. Chitosan induced the accumulation of H_2O_2 in tomatoes [127]. Upon pathogen infection, one of the quickest responses is an increase in cytosolic Ca^{2+} [128]. Within 20 min of being treated with chitosan, Glycine max suspension-cultured cells began to synthesize callose. However, in the absence of exogenous Ca^{2+}, chitosan-induced callose production was not achievable [72]. Chitosan increased the amount of free cytosolic Ca^{2+} in Arabidopsis and stomatal closure [129].

NO, another messenger, implicated in the plant defense response against pathogens and involved in chitosan-induced resistance [114]. Downy mildew-infected pearl millet seedlings treated with chitosan exhibited increased NO buildup commencing 2 h after inoculation, as well as protection from downy mildew [114]. Chitosan induced the generation of NO and phosphatidic acid in tomato cell culture while the phospholipase-mediated signaling pathway was inhibited after using NO scavenger, indicating that the production of phosphatidic acid during the plant defensive response needed NO [130].

Plant hormones regulate growth processes in plants and play important roles in plant responses to biotic and abiotic stresses [131]. Salicylic acid, jasmonic acid, and ethylene play a crucial role as signaling molecules in modulating Arabidopsis' responses to biotic and abiotic stress. Jasmonic acid and ethylene are central signaling molecules

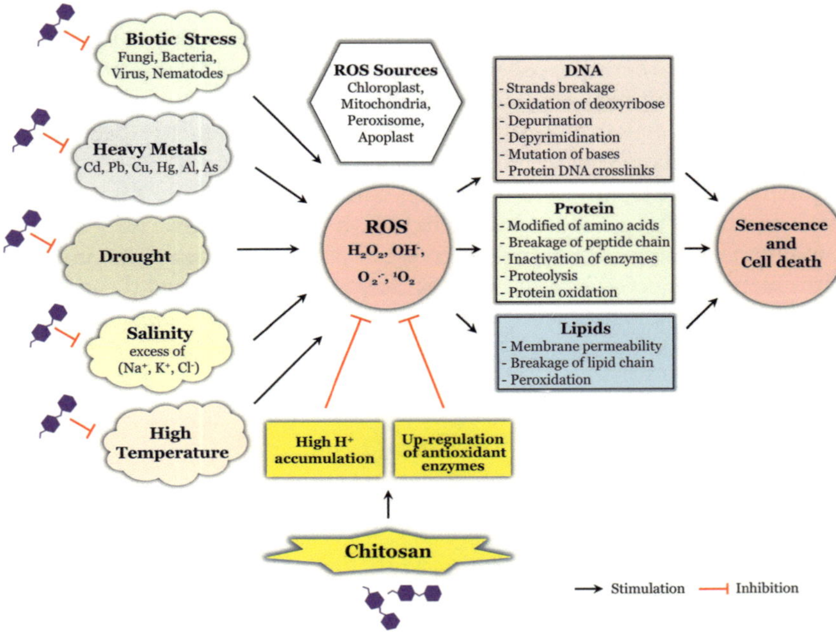

Figure 5.
An overview of how chitosan helps plants withstand phytopathogens, abiotic stress, and ROS.

in the induced systemic resistance; on the other hand, salicylic acid is involved in systemic acquired resistance, which occurs when a pathogenic attack on one area of the plant results in resistance in other parts. It has been suggested that jasmonic acid is a component of a signal transduction pathway that controls the activation of genes involved in plant defensive responses to pathogen invasion. Chitosan significantly increased the jasmonic acid in wounded rice leaves [132]. ABA regulates the intensity and speed of callose deposition. Additionally, ABA-mediated signaling transduction is crucial for plants to respond to abiotic and biotic stresses [133]. Chitosan activated the defense signaling pathways in tomato plants against *A. solani* and *X. vesicatoria* [26].

5.5.2 Activation of symbiotic signaling

Beneficial microorganisms are able to form a symbiotic connection while living in close proximity to their plant hosts, which aids in the acquisition of nutrients necessary for plant growth. Additionally, the plant meets some of the requirements that helpful microorganisms need to complete their metabolic processes for growth and reproduction [134]. Chitosan triggered symbiotic signaling between plants and beneficial microbes [135]. Lipo-chitooligosaccharides, which are chitin oligosaccharides bound to a lipid moiety, have been discovered to be released by mycorrhizal fungi and rhizobium bacteria during the development of symbiotic interactions and have been recognized as key signaling molecules triggering the plant symbiotic response [136]. Chitosan undergoes enzymatic degradation without harming the beneficial rhizosphere biota in the soil and promotes symbiotic interactions between plants and microorganisms, causing changes in the rhizosphere's microbial balance and harming plant diseases [44].

5.6 Usage as bio-fertilizer

As a result of the repeated use of inorganic fertilizers that are difficult to decompose, the toxicity of the soil increased, which affected the beneficial microorganisms present in the soil and the properties of the soil. Therefore, the use of chitosan at low concentrations as a bio-fertilizer was a safe and effective alternative to avoid the risks of using inorganic fertilizers. Utilizing chitosan as a biofertilizer showed a significant decrease in late blight infestation of potato tubers and an increase in plant nutrient uptake [137]. The addition of chitosan to soil improved the phosphorous and nitrogen content in *Eustoma grandiflorum* [68].

Large amounts of inorganic fertilizers are lost in the soil due to the inability of plants to absorb them, causing farmers to over-apply them, resulting in their presence in the soil at higher than required rates, causing soil toxicity, water pollution, and damage to crops, particularly vegetables, which are severely affected by fertilizer toxicity [138]. However, when chitosan coating on fertilizer is added to the soil, it improves the absorption of inorganic fertilizer by plants, which minimizes the use of fertilizers, makes the soil less toxic, and reduces the production cost [139].

6. Recovery of contaminated agricultural wastewater and soil

The concentration of heavy metals and other contaminants in the environment is increasing rapidly as a result of multiple human activities. Heavy metals are dangerous because they are highly poisonous, do not biodegrade, and cause cancer and other disorders. Therefore, it is necessary to find an efficient way to remove them from the environment and dispose of them. From this perspective, bio adsorption is acknowledged as an affordable and effective solution [140]. Effective pollutant removal makes it possible to reuse valuable resources like cultivable soil and fresh water. The application of chitosan-based adsorbents in these areas has been extensively studied due to their low production costs, biocompatible and biodegradable nature, strong resistance to antimicrobial attack, and absence of the creation of potentially toxic secondary end products [140]. The chemical composition of chitosan makes it simple to combine with certain ions, molecules, and other compounds to produce complex structures for specific applications. Carboxylated graphene oxide-chitosan (GO-COOH/CS) spheres were used for the immobilization of Cu^{2+} from water and soil, as well as for reducing the bioaccumulation of Cu^{2+} from wheat plants [141]. Cesium-contaminated clay can be cleaned using ionized chitosan and magnetic microgels functionalized with Prussian blue analogs, 200 mg/g of ionized chitosan hydrochloride can achieve 87.6% cesium release from clay in about 2 h [142].

A combination of chitosan and duckweed was assessed for its ability to remove boron from water. Chitosan beads had the maximum boron absorption capacity of chitosan at 3.18 mg/g [143]. Chitosan can help reduce the environmental effects of industrial wastewater treatments and soil acidity by reducing CO_2 and SO_2 [144]. Columns packed with chitosan have the ability to remove arsenic from groundwater [145]. A composite consisting of chitosan and hyacinth extract effectively absorbs Cu, Pb, and Cd ions from water [146]. Cd removal from the soil and aquatic environments was examined using the magnesium oxide biochar-chitosan composite (MgO-BCR-W) [147].

The adsorbent made of chitosan/MnO_2 nanocomposite was employed to extract Cr (VI) from the aqueous solutions [148]. A magnetic chitosan/polyacrylic acid nanocomposite successfully adsorbed Pb (II) from an aqueous solution [149]. Iron chitosan microspheres were synthesized by ionotropic gelation for the removal of arsenic (V) from water [150]. Chitosan contains functional amino and hydroxyl groups, which enables it to form compounds with heavy metals. Foliar application of chitosan can reduce the harmful effects of cadmium in *Brassica rapa* L. [151]. Herbicides can be made more effective by adding chitosan to formulations, which lowers the amount utilized and the risk of hazard accumulation in the environment. A formulation consisting of chitosan and glyphosate exhibits lower phytotoxicity and higher herbicidal efficacy and releases active substances better than using glyphosate alone [152]. Chitosan and tripolyphosphate nanoparticles are efficient carriers to reduce soil sorption, cytotoxicity, and mutagenicity of paraquat and enhance their herbicide activity [153].

7. Chitosan future prospect

Despite extensive research on chitosan, the mechanism of action of chitosan in regulating plant immunity and suppressing the pathogen has not been sufficiently elucidated. It is thought that the mode of action of chitosan may be more complex and involve a series of overlapping details that need to be further studied in the future. Proteomics is one of the contemporary disciplines that has been successfully used to investigate the global variations in protein expression in biological organisms under a variety of environmental conditions [154]. There are a lot of proteomic studies that explain the chitosan mode of action in withstanding biotic and abiotic stresses in plants. The inhibitory effect of chitosan on *P. expansum* was proteomic analyzed, and 26 proteins were identified and grouped according to their potential biological roles [155]. A comprehensive proteomic study of chitosan-responsive proteins explained the inhibitory mechanism of chitosan against *F. oxysporum* f. sp. *cucumerinum*. This led to the identification of 62 expressed proteins involved in the hindering of the Fusarium cell wall, disrupting DNA, and disrupting structural and functional protein biosynthesis and explained how chitosan influences metabolic pathways [31]. We wish plants and pathogens treated with chitosan would receive abundant proteomic studies in order to make maximum use of chitosan in sustainable agriculture.

Acknowledgements

The authors would like to thank the staff members of Plant Pathology Research Institute, Agricultural Research Center, Giza, Egypt.

Conflict of interest

The authors declare no conflict of interest.

Author details

Magdi A.E. Abdellatef*†, Eman Elagamey*† and Said M. Kamel
Plant Pathology Research Institute, Agricultural Research Center (ARC), Giza, Egypt

*Address all correspondence to: magdi_abdellatef@yahoo.com
and emanelagamey@yahoo.com

† These authors contributed equally.

References

[1] No HK, Park NY, Lee SH, Meyers SP. Antibacterial activity of chitosans and chitosan oligomers with different molecular weights. International Journal of Food Microbiology. 2002;**74**(1-2):65-72. DOI: 10.1016/S0168-1605(01)00717-6

[2] Ghormade V, Pathan E, Deshpande M. Can fungi compete with marine sources for chitosan production? International Journal of Biological Macromolecules. 2017;**104**:1415-1421. DOI: 10.1016/j.ijbiomac.2017.01.112

[3] Mujtaba M, Khawar KM, Camara MC, Carvalho LB, Fraceto LF, Morsi RE, et al. Chitosan-based delivery systems for plants: A brief overview of recent advances and future directions. International Journal of Biological Macromolecules. 2020;**154**:683-697. DOI: 10.1016/j.ijbiomac.2020.03.128

[4] Chirkov SN. The antiviral activity of chitosan (review). Applied Biochemistry and Microbiology. 2002;**38**:1-8. DOI: 10.1023/A:1013206517442

[5] Kulikov SN, Chirkov SN, Il'ina AV, Lopatin SA, Varlamov VP. Effect of the molecular weight of chitosan on its antiviral activity in plants. Applied Biochemistry and Microbiology. 2006;**42**(2):200-203. DOI: 10.1134/S0003683806020165

[6] Zhao J, Pan L, Zhou M, Yang Z, Meng Y, Zhang X. Comparative physiological and transcriptomic analyses reveal mechanisms of improved osmotic stress tolerance in annual ryegrass by exogenous chitosan. Genes. 2019;**10**(11):853. DOI: 10.3390/genes10110853

[7] Ahmed KBM, Khan MMA, Siddiqui H, Jahan A. Chitosan and its oligosaccharides, a promising option for sustainable crop production—A review. Carbohydrate Polymers. 2020;**227**:115331. DOI: 10.1016/j.carbpol.2019.115331

[8] Hudson SM, Jenkins DW. Chitin and Chitosan, Encyclopedia of Polymer Science and Technology. NJ: Wiley Interscience; 2001. DOI: 10.1002/0471440264.pst052

[9] Rahman L, Goswami J. Recent development on physical and biological properties of chitosan-based composite films with natural extracts: A review. The Journal of Bioactive and Compatible Polymers. 2021;**36**(3):225-236. DOI: 10.1177/08839115211014218

[10] Dash M, Chiellini F, Ottenbrite RM, Chiellini E. Chitosan—A versatile semi-synthetic polymer in biomedical applications. Progress in Polymer Science. 2011;**36**:981-1014. DOI: 10.1016/j.progpolymsci.2011.02.001

[11] Aranaz I, Alcántara AR, Civera MC, Arias C, Elorza B, Heras Caballero A, et al. Chitosan: An overview of its properties and applications. Polymers-Basel. 2021;**13**:3256. DOI: 10.3390/polym13193256

[12] Rinaudo M. Chitin and Chitosan: Properties and applications. Progress in Polymer Science. 2006;**31**(7):603-632. DOI: 10.1016/j.progpolymsci.2006.06.001

[13] Ke CL, Deng FS, Chuang CY, Lin CH. Antimicrobial actions and applications of chitosan. Polymers. 2021;**13**(6):904. DOI: 10.3390/polym13060904

[14] Chakraborty M, Hasanuzzaman M, Rahman M, Khan MAR, Bhowmik P, Mahmud NU, et al. Mechanism of plant growth promotion

and disease suppression by chitosan biopolymer. Agriculture. 2020;**10**(12):624. DOI: 10.3390/agriculture10120624

[15] Palma-Guerrero J, Lopez-Jimenez JA, Pérez-Berná AJ, Huang IC, Jansson HB, Salinas J, et al. Membrane fluidity determines sensitivity of filamentous fungi to chitosan. Molecular Microbiology. 2010;**75**(4):1021-1032. DOI: 10.1111/j.1365-2958.2009.07039.x

[16] Ma Z, Garrido-Maestu A, Jeong KC. Application, mode of action, and in vivo activity of chitosan and its microand nanoparticles as antimicrobial agents: A review. Carbohydrate Polymers. 2017;**176**:257-265. DOI: 10.1016/j.carbpol.2017.08.082

[17] Divya K, Vijayan S, George TK, Jisha MS. Antimicrobial properties of chitosan nanoparticles: Mode of action and factors affecting activity. Fibres and Polymers. 2017;**18**(2):221-230. DOI: 10.1007/s12221-017-6690-1

[18] Wang X, Du Y, Fan L, Liu H, Hu Y. Chitosan-metal complexes as antimicrobial agent: Synthesis, characterization and structure-activity study. Polymer Bulletin. 2005;**55**(1):105-113. DOI: 10.1007/s00289-005-0414-1

[19] Goy RC, Britto DD, Assis OB. A review of the antimicrobial activity of chitosan. Polímeros. 2009;**19**:241-247. DOI: F10.1590/S0104-14282009000300013

[20] Yan D, Li Y, Liu Y, Li N, Zhang X, Yan C. Antimicrobial properties of Chitosan and Chitosan derivatives in the treatment of enteric infections. Mol. 2021;**26**(23):7136. DOI: 10.3390/molecules26237136

[21] Hosseinnejad M, Jafari SM. Evaluation of different factors affecting antimicrobial properties of chitosan. International Journal of Biological Macromolecules. 2016;**85**:467-475. DOI: 10.1016/j.ijbiomac.2016.01.022

[22] Li YC, Sun XJ, Bi Y, Ge YH, Wang Y. Antifungal activity of Chitosan on *fusarium sulphureum* in relation to dry rot of potato tuber. Agricultural Sciences in China. 2009;**8**:597-604. DOI: 10.1016/S1671-2927(08)60251-5

[23] Hernandez-Lauzardo AN, Bautista-Banos S, Velazquez-del Valle MG, Mendez-Montealvo MG, Anchez-Rivera MM, Bello-Perez LA. Antifungal effects of chitosan with different molecular weights on *In vitro* development of *Rhizopus stolonifer* (Ehrenb.:Fr.) Vuill. Carbohydrate Polymers. 2008;**73**:541-547. DOI: 10.1016/j.carbpol.2007.12.020

[24] De Oliveira Jr EN, De Melo IS, Franco TT. Changes in hyphal morphology due to chitosan treatment in some fungal species. Brazilian Archives of Biology and Technology. 2012;**55**:637-646. DOI: 10.1590/S1516-89132012000500001

[25] Guo Z, Chen R, Xing R, Liu S, Yu H, et al. Novel derivatives of chitosan and their antifungal activities *in vitro*. Carbohydrate Research. 2006;**341**:351-354. DOI: 10.1016/j.carres.2005.11.002

[26] Ramkissoon A, Francis J, Bowrin V, Ramjegathesh R, Ramsubhag A, Jayaraman J. Bio efficacy of a chitosan based elicitor on *A. solani* and *Xanthomonas vesicatoria* infections in tomato under tropical conditions. The Annals of Applied Biology. 2016;**169**(2):274-283. DOI: 10.1111/aab.12299

[27] Al-Hetar MY, Zainal Abidin MA, Sariah M, Wong MY. Antifungal activity of chitosan against *fusarium oxysporum*

f. sp. *Cubense*. Journal of Applied Polymer Science. 2011;**120**:2434-2439. DOI: 10.1002/app.33455

[28] Prapagdee B, Kotchadat K, Kumsopa A, Visarathanonth N. The role of chitosan in protection of soybean from sudden death syndrome caused by *fusarium solani* f. sp. *glycines*. Bioresource Technology. 2007;**98**:1353-1358. DOI: 10.1016/j.biortech.2006.05.029

[29] Rabea EI, Steurbaut W. Chemically modified chitosans as antimicrobial agents against some plant pathogenic bacteria and fungi. Plant Protection Science. 2010;**46**(4):149-158. DOI: 10.17221/9/2009-PPS

[30] El Hassni M, El Hadrami A, Daayf F, El BEA, Hadrami I. Chitosan, antifungal product against *fusarium oxysporum* f. sp. *albedinis* and elicitor of defence reactions in date palm roots. Phytopathologia Mediterranea. 2004;**43**:195-204. DOI: 10.14601/Phytopathol_Mediterr-1743

[31] Elagamey E, Abdellatef MAE, Arafat MY. Proteomic insights of chitosan mediated inhibition of *Fusarium oxysporum* f. sp. *cucumerinum*. Journal of Proteomics. 2022;**260**:104560. DOI: 10.1016/j.jprot.2022.104560

[32] Coutinho TC, Ferreira MC, Rosa LH, de Oliveira AM, Júnior ENO. *Penicillium citrinum* and *penicillium mallochii*: New phytopathogens of orange fruit and their control using chitosan. Carbohydrate Polymers. 2020;**234**:115918. DOI: 10.1016/j.carbpol.2020.115918

[33] Akter J, Jannat R, Hossain MM, Ahmed JU, Rubayet MT. Chitosan for plant growth promotion and disease suppression against anthracnose in chilli. International Journal of Agriculture Environment and Biotechnology. 2018;**3**:806-817. DOI: 10.22161/ijeab/3.3.13

[34] Li SJ, Zhu TH. Biochemical response and induced resistance against anthracnose (*Colletotrichum camelliae*) of camellia (*Camellia pitardii*) by chitosan oligosaccharide application. Forest Pathology. 2013;**43**:67-76. DOI: 10.1111/j.1439-0329.2012.00797.x

[35] Raafat D, von Bargen K, Haas A, Sahl HG. Insights into the mode of action of chitosan as an antibacterial compound. Applied and Environmental Microbiology. 2008;**74**(12):3764-3773. DOI: 10.1128/AEM.00453-08

[36] Beausejour J, Clermont N, Beaulieu C. Effect of *Streptomyces melanosporofaciens* strain EF-76 and of chitosan on common scab of potato. Plant and Soil. 2003;**256**:463-468. DOI: 10.1023/A:1026177714855

[37] Li B, Liu B, Shan C, Ibrahim M, Lou Y, Wang Y, et al. Antibacterial activity of two chitosan solutions and their effect on rice bacterial leaf blight and leaf streak. Pest Management Science. 2013;**69**:312-320. DOI: 10.1002/ps.3399

[38] Li B, Liu B, Su T, Fang Y, Xie G, Wang G, et al. Effect of chitosan solution on the inhibition of *Pseudomonas fluorescens* causing bacterial head rot of broccoli. Plant Pathology Journal. 2010;**26**:189-193. DOI: 10.5423/PPJ.2010.26.2.189

[39] Yang C, Li B, Ge M, Zhou K, Wang Y, Luo J, et al. Inhibitory effect and mode of action of chitosan solution against rice bacterial brown stripe pathogen *Acidovorax avenae* subsp. *avenae* RS-1. Carbohydrate Research. 2014;**4**(391):48-54. DOI: 10.1016/j.carres.2014.02.025

[40] Li B, Shi S, Shan C, Zhou Q, Ibrahim M, Wang Y, et al. Effect of chitosan solution on the inhibition of *Acidovorax citrulli* causing bacterial

fruit blotch of watermelon. Journal of the Science of Food and Agriculture. 2013b;**93**:1010-1015. DOI: 10.1002/jsfa.5812

[41] Jia X, Meng Q, Zeng H, Wang W, Yin H. Chitosan oligosaccharide induces resistance to tobacco mosaic virus in Arabidopsis via the salicylic acid-mediated signaling pathway. Scientific Reports. 2016;**18**(6):26144. DOI: 10.1038/srep26144

[42] Abdelkhalek A, Qari SH, Abu-Saied MAA-R, Khalil AM, Younes HA, Nehela Y, et al. Chitosan nanoparticles inactivate alfalfa mosaic virus replication and boost innate immunity in *Nicotiana glutinosa* plants. Plants. 2021;**10**(12):2701. DOI: 10.3390/plants10122701

[43] El-Ansary MSM, Khalifa EZ, Hemdan SM. Influence of fungal Chitosan to control root-knot nematode *Meloidogyne incognita* on Banana plants. The Egyptian Journal of Phytopathology. 2013;**41**(1):43-52. DOI: 10.21608/ejp.2013.101967

[44] Escudero N, Lopez-Moya F, Ghahremani Z, Zavala-Gonzalez EA, Alaguero-Cordovilla A, Ros-Ibañez C, et al. Chitosan increases tomato root colonization by *Pochonia chlamydosporia* and their combination reduces root-knot nematode damage. Frontiers in Plant Science. 2017;**8**:1415. DOI: 10.3389/fpls.2017.01415

[45] Khan A, Tariq M, Ahmad F, Mennan S, Khan F, Asif M, et al. Assessment of nematicidal efficacy of chitosan in combination with botanicals against *Meloidogyne incognita* on carrot. Acta Agriculturæ Scandinavica Section B. 2021;**71**(4):225-236. DOI: 10.1080/09064710.2021.1880620

[46] Singh RR, Chinnasri B, De Smet L, Haeck A, Demeestere K, Van Cutsem P, et al. Systemic defense activation by COS-OGA in rice against root-knot nematodes depends on stimulation of the phenylpropanoid pathway. Plant Physiology and Biochemistry. 2019;**1**(142):202-210. DOI: 10.1016/j.plaphy.2019.07.003

[47] Bistgani ZE, Siadat SA, Bakhshandeh A, Pirbalouti AG, Hashemi M. Interactive effects of drought stress and chitosan application on physiological characteristics and essential oil yield of *Thymus daenensis* Celak. The Crop Journal. 2017;**5**(5):407-415. DOI: 10.1016/j.cj.2017.04.003

[48] Jiao Z, Li Y, Li J, Xu X, Li H, Lu D, et al. Effects of exogenous chitosan on physiological characteristics of potato seedlings under drought stress and rehydration. Potato Research. 2012;**55**(3):293-301. DOI: 10.1007/s11540-012-9223-8

[49] Yang F, Hu J, Li J, Wu X, Qian Y. Chitosan enhances leaf membrane stability and antioxidant enzyme activities in apple seedlings under drought stress. Plant Growth Regulation. 2009;**58**(2):131-136. DOI: 10.1007/s10725-009-9361-4

[50] Farouk S, Amany AR. Improving growth and yield of cowpea by foliar application of chitosan under water stress. Egyptian Journal of Biology. 2012;**2**(10):1341-1358. DOI: 10.21608/JPP.2011.85662

[51] Pirbalouti AG, Malekpoor F, Salimi A, Golparvar A. Exogenous application of chitosan on biochemical and physiological characteristics, phenolic content and antioxidant activity of two species of basil (*Ocimum ciliatum* and *Ocimum basilicum*) under reduced irrigation. Scientia Horticulturae. 2017;**217**:114-122. DOI: 10.1016/j.scienta.2017.01.031

[52] Dzung NA, Khanh VTP, Dzung TT. Research on impact of chitosan oligomers on biophysical characteristics, growth, development and drought resistance of coffee. Carbohydrate Polymers. 2011;**84**:751-755. DOI: 10.1016/J. CARBPOL.2010.07.066Corpus ID: 85289773

[53] Bittelli M, Flury M, Campbell GS, Nichols EJ. Reduction of transpiration through foliar application of chitosan. Agricultural and Forest Meteorology. 2001;**107**:167-175. DOI: 10.1016/ S0168-1923(00)00242-2

[54] Jabeen N, Ahmad R. The activity of antioxidant enzymes in response to salt stress in safflower (*Carthamus tinctorius* L.) and sunflower (*Helianthus annuus* L.) seedlings raised from seed treated with chitosan. Journal of the Science of Food and Agriculture. 2013;**93**(7):1699-1705. DOI: 10.1002/jsfa.5953

[55] Ma L, Li Y, Yu C, Wang Y, Li X, Li N, et al. Alleviation of exogenous oligochitosan on wheat seedlings growth under salt stress. Protoplasma. 2012;**249**(2):393-399. DOI: 10.1007/ s00709-011-0290-5

[56] Al-Tawaha AR, Turk MA, Al-Tawaha AR, Alu'datt MH, Wedyan M, Al-Ramamneh EA, Hoang AT. Using chitosan to improve growth of maize cultivars under salinity conditions. Bulgarian Journal of Agricultural Science. 2018;**24**(3):437-442

[57] González LM, Guerrero YR, Rodríguez A, Vázquez MN. Effect of seed treatment with chitosan on the growth of rice (*Oryza sativa* L.) seedlings cv. INCA LP-5 in saline medium. Cultivos Tropicales. 2015;**36**(1):143-150

[58] Yahyaabadi HM, Asgharipour MR, Basiri M. Role of chitosan in improving salinity resistance through some morphological and physiological characteristics in fenugreek (*Trigonella foenum-graecum* L.). Journal of Science and Technology of Greenhouse Culture. 2016;**7**(25):165-174. DOI: 10.18869/ acadpub.ejgcst.7.1.165

[59] Ibrahim EA, Ramadan WA. Effect of zinc foliar spray alone and combined with humic acid or/and chitosan on growth, nutrient elements content and yield of dry bean (*Phaseolus vulgaris* L.) plants sown at different dates. Scientia Horticulturae. 2015;**184**:101-105. DOI: 10.1016/j.scienta.2014.11.010

[60] Xue GX, Gao HY, Li PM, Zou Q. Effects of chitosan treatment on physiological and biochemical characteristics in cucumber seedlings under low temperature. Zhi wu Sheng li yu fen zi Ssheng wu xue xue bao= Journal of Plant Physiology and Molecular Biology. 2004;**30**(4):441-448

[61] Guan Y, Hu J, Wang X, Shao C. Seed priming with chitosan improves maize germination and seedling growth in relation to physiological changes under low temperature stress. Journal of Zhejiang University. Science. B. 2009;**10**:427-433. DOI: 10.1631/jzus.B0820373

[62] Sheikha SA, Al-Malki FM. Growth and chlorophyll responses of bean plants to the chitosan applications. European Journal of Scientific Research. 2011;**50**(1):124-134

[63] Chibu H, Shibayama H, Arima S. Effects of chitosan application on the shoot growth of rice and soybean. Japanese Journal of Crop Science. 2002;**71**(2):206-211. DOI: 10.1626/jcs.71.206

[64] Shehata SA, Fawzy ZF, El-Ramady HR. Response of cucumber plants to foliar application of chitosan and yeast under greenhouse conditions. Australian Journal of Applied Science. 2012;**6**(4):63-71

[65] Mondal MMA, Malek MA, Puteh AB, Ismail MR, Ashrafuzzaman M, Naher L. Effect of foliar application of chitosan on growth and yield in okra. Australian Journal of Crop Science. 2012;**6**(5):918-921

[66] Sultana S, Islam M, Khatun MA, Hassain M, Huque R. Effect of foliar application of oligo-chitosan on growth, yield and quality of tomato and eggplant. Asian Journal of Advances in Agricultural Research. 2017;**11**(2):36-42. DOI: 10.3923/ajar.2017.36.42

[67] Salachna P, Byczyńska A, Jeziorska I, Udycz E. Plant growth of *Verbena bonariensis* L. after chitosan, gellan gum or iota-carrageenan foliar applications. World Science News. 2017;**62**:111-123

[68] Ohta K, Atarashi H, Shimatani Y, Matsumoto S, Asao T, Hosoki T. Effects of chitosan with or without nitrogen treatments on seedling growth in *Eustoma grandiflorum* (Raf.) Shinn. cv. Kairyou Wakamurasaki. Journal of the Japanese Society for Horticultural Science. 2000;**69**(1):63-65. DOI: 10.2503/jjshs.69.63

[69] Khan W, Prithiviraj B, Smith DL. Chitosan and chitin oligomers increase phenylalanine ammonia-lyase and tyrosine ammonia-lyase activities in soybean leaves. Journal of Plant Physiology. 2003;**160**(8):859-863. DOI: 10.1078/0176-1617-00905

[70] Narula K, Elagamey E, Abdellatef MAE, Sinha A, Ghosh S, Chakraborty N, et al. Chitosan-triggered immunity to Fusarium in chickpea is associated with changes in the plant extracellular matrix architecture, stomatal closure and remodelling of the plant metabolome and proteome. The Plant Journal. 2020;**103**(2):561-583. DOI: 10.1111/tpj.14750

[71] Iriti M, Faoro F. Chitosan as a MAMP, searching for a PRR. Plant Signaling & Behavior. 2009;**4**(1):66-68. DOI: 10.4161/psb.4.1.7408

[72] Köhle H, Jeblick W, Poten F, Blaschek W, Kauss H. Chitosan-elicited callose synthesis in soybean cells as a Ca^{2+}-dependent process. Plant Physiology. 1985;**77**(3):544-551. DOI: 10.1104/pp.77.3.544

[73] Gow NAR, Latge JP, Munro CA. The fungal cell wall: Structure, biosynthesis, and function. Microbiology Spectrum. 2017;**5**(3):5-3. DOI: 10.1128/microbiolspec.FUNK-0035-2016

[74] Benhamou N. Ultrastructural and cytochemical aspects of chitosan on *fusarium oxysporum* f. sp. *radicis-lycopersici*, agent of tomato crown and root rot. Phytopathology. 1992;**82**:1185-1193. DOI: 10.1016/S0885-5765(05)80098-0

[75] Ait Barka E, Eullaffroy P, Clément C, Vernet G. Chitosan improves development, and protects *Vitis vinifera* L. against *Botrytis cinerea*. Plant Cell Reports. 2004;**22**:608-614. DOI: 10.1007/s00299-003-0733-3

[76] Baños SB, López MH, Molina EB. Growth inhibition of selected fungi by chitosan and plant extracts. Revista Mexicana de Fitopatología. 2004;**22**(2):178-186

[77] Pasquina-Lemonche L, Burns J, Turner RD, Kumar S, Tank R, Mullin N, et al. The architecture of the gram-positive bacterial cell wall. Nature. 2020;**582**(7811):294-297. DOI: 10.1038/s41586-020-2236-6

[78] Cheung RC, Ng TB, Wong JH, Chan WY. Chitosan: An update on potential biomedical and pharmaceutical applications. Marine Drugs. 2015;**13**:5156-5186. DOI: 10.3390/md13085156

[79] Chung YC, Su YP, Chen CC, Jia G, Wang HL, Wu JG, et al. Relationship

between antibacterial activity of chitosan and surface characteristics of cell wall. Acta Pharmacologica Sinica. 2004;**25**(7):932-936

[80] Tan H, Ma R, Lin C, Liu Z, Tang T. Quaternized chitosan as an antimicrobial agent: Antimicrobial activity, mechanism of action and biomedical applications in orthopedics. International Journal of Molecular Sciences. 2013;**14**(1):1854-1869. DOI: 10.3390/ijms14011854

[81] Farag SMA, Elhalag KMA, Mohamed H, Hagag MH, Khairy ASM, Ibrahim HM, et al. Potato bacterial wilt suppression and plant health improvement after application of different antioxidants. Journal of Phytopathology. 2017;**65**:522-537. DOI: 10.1111/jph.12589

[82] Li J, Wu Y, Zhao L. Antibacterial activity and mechanism of chitosan with ultra-high molecular weight. Carbohydrate Polymers. 2016;**148**:200-205. DOI: 10.1016/j.carbpol.2016.04.025

[83] Annaian S, Kandasamy K, Lakshman N. Preparation, characterization and antibacterial activity of chitosan and phosphorylated chitosan from cuttlebone of *Sepia kobiensis*. Biotechnology Reports. 2016;**9**:25-30. DOI: 10.1016/j.btre.2015.10.007

[84] Sapers GM. Chitosan enhances control of enzymatic browning in apple and pear juice by filtration. Journal of Food Science. 1992;**57**(5):1192-1193. DOI: 10.1111/j.1365-2621.1992.tb11296.x

[85] Chung YC, Wang HL, Chen YM, Li SL. Effect of abiotic factors on the antibacterial activity of chitosan against waterborne pathogens. Bioresource Technology. 2003;**88**:179-184. DOI: 10.1016/s0960-8524(03)00002-6

[86] Rabea EI, Badawy ME, Stevens CV, Smagghe G, Steurbaut W. Chitosan as antimicrobial agent: Applications and mode of action. Biomacromolecules. 2003;**4**(6):1457-1465. DOI: 10.1021/bm034130m

[87] Goy RC, Morais ST, Assis OB. Evaluation of the antimicrobial activity of chitosan and its quaternized derivative on *E. coli* and *S. aureus* growth. Revista Brasileira de Farmacognosia. 2016;**26**:122-127. DOI: 10.1016/j.bjp.2015.09.010

[88] Liang C, Yuan F, Liu F, Wang Y, Gao Y. Structure and antimicrobial mechanism of ε-polylysine–chitosan conjugates through Maillard reaction. International Journal of Biological Macromolecules. 2014;**70**:427-434. DOI: 10.1016/j.ijbiomac.2014.07.012

[89] Acar O, Aki C, Erdugan H. Fungal and bacterial diseases control with Elexa TM plant booster. Fresenius Environmental Bulletin. 2008;**17**:797-802

[90] Li B, Wang X, Chen RX, Huangfu WG, Xie GL. Antibacterial activity of chitosan solution against *Xanthomonas* pathogenic bacteria isolated from *Euphorbia pulcherrima*. Carbohydrate Polymers. 2008;**72**:287-292. DOI: 10.1016/j.carbpol.2007.08.012

[91] Dong OX, Ronald PC. Genetic engineering for disease resistance in plants: Recent progress and future perspectives. Plant Physiology. 2019;**180**:26-38. DOI: 10.1104/pp.18.01224

[92] Mansouri S, Lavigne P, Corsi K, Benderdour M, Beaumont E, Fernandes JC. Chitosan-DNA nanoparticles as non-viral vectors in gene therapy strategies to improve transfection efficacy. European Journal of Pharmaceutics and Biopharmaceutics. 2004;**57**:1-8. DOI: 10.1016/s0939-6411(03)00155-3

[93] Akhter G, Khan TA. Evaluation of some plant extracts for nemato-toxic

potential against juveniles of *Meloidogyne incognita in vitro*. The Journal of Phytopharmacology. 2018;7(2):141-145. DOI: 10.31254/phyto.2018.7207

[94] Pichyangkura R, Chadchawan S. Biostimulant activity of chitosan in horticulture. Scientia Horticulturae. 2015;**196**:49-65. DOI: 10.1016/j.scienta.2015.09.031

[95] Din J, Khan SU, Ali I, Gurmani AR. Physiological and agronomic response of canola varieties to drought stress. The Journal of Animal and Plant Sciences. 2011;**21**(1):78-82. DOI: 10.4081/ijas.2011.e11

[96] Li Q, Zhang X, Lv Q, Zhu D, Qiu T, Xu Y, et al. Physcomitrella Patens dehydrins (PpDHNA and PpDHNC) confer salinity and drought tolerance to transgenic Arabidopsis plants. Frontiers in Plant Science. 2017;**8**:1316. DOI: 10.3389/fpls.2017.01316

[97] Hidangmayum A, Dwivedi P, Katiyar D, Hemantaranjan A. Application of chitosan on plant responses with special reference to abiotic stress. Physiology and Molecular Biology of Plants. 2019;**25**(2):313-326. DOI: 10.1007/s12298-018-0633-1

[98] Nazarli H, Faraji F, Zardashti MR. Effect of drought stress and polymer on osmotic adjustment and photosynthetic pigments of sunflower. Cercetari Agronomice in Moldova. 2011;**1**(145):35-41

[99] Lai QX, Bao ZY, Zhu ZJ, Qian QQ, Mao BZ. Effects of osmotic stress on antioxidant enzymes activities in leaf discs of PSAG12-IPT modified Gerbera. Journal of Zhejiang University. Science B. 2007;**8**(7):458-464. DOI: 10.1631/jzus.2007.B0458

[100] Khan WM, Prithiviraj B, Smiyh DL. Effect of foliar application of chitinoligosaccharides on photosynthesis of maize and soybean. Photosynthetica. 2002;**40**(4):621-624. DOI: 10.1023/A:1024320606812

[101] Meloni DA, Oliva MA, Martinez CA, Cambraia J. Photosynthesis and activity of superoxide dismutase, peroxidase and glutathione reductase in cotton under salt stress. Environmental and Experimental Botany. 2003;**49**(1):69-76. DOI: 10.1016/S0098-8472(02)00058-8

[102] Sun Z, Shi C, Wang X, Fang Q, Huang J. Synthesis, characterization, and antimicrobial activities of sulfonated chitosan. Carbohydrate Polymers. 2017;**2**(155):321-328

[103] Khan MN, Siddiqui MH, Mohammad F, Naeem M, Khan MMA. Calcium chloride and gibberellic acid protect linseed (*Linum usitatissimum* L.) from NaCl stress by inducing antioxidative defence system and osmoprotectant accumulation. Acta Physiologiae Plantarum. 2010;**32**:121-132. DOI: 10.1007/s11738-009-0387-z

[104] Choi YS, Kim YM, Hwang OJ, Han YJ, Kim SY, Kim JI. Overexpression of Arabidopsis ABF3 gene confers enhanced controlled-release and water-retention. Carbohydrate Polymers. 2008;**72**:240-247. DOI: 10.1007/s11816-012-0245-0

[105] Torres-Rodriguez JA, Reyes-Perez JJ, Castellanos T, Angulo C, Quinones-AguilarEE,Hernandez-MontielLG. A biopolymer with antimicrobial properties and plant resistance inducer against phytopathogens: Chitosan. Not Bot Horti Agrobo. 2021;**49**(1):12231. DOI: 10.15835/nbha49112231

[106] Hameed A, Sheikh M, Hameed A, Farooq T, Basra S, Jamil A. Chitosan priming enhances the seed germination, antioxidants, hydrolytic enzymes,

soluble proteins and sugars in wheat seeds. Agrochimica. 2013;**67**:32-46

[107] Ziani K, Ursúa B, Maté JI. Application of bioactive coatings based on chitosan for artichoke seed protection. Crop Protection. 2010;**29**(8):853-859. DOI: 10.1016/j.cropro.2010.03.002

[108] Saavedra GM, Figueroa NE, Poblete LA, Cherian S, Figueroa CR. Effects of preharvest applications of methyl jasmonate and chitosan on postharvest decay, quality and chemical attributes of Fragaria chiloensis fruit. Food Chemistry. 2016;**1**(190):448-453. DOI: 10.1016/j.foodchem.2015.05.107

[109] Kowalski B, Terry FJ, Herrera L, Peñalver DA. Application of soluble chitosan *in vitro* and in the greenhouse to increase yield and seed quality of potato minitubers. Potato Research. 2006;**49**:167-176. DOI: 10.1007/s11540-006-9015-0

[110] González Gómez H, Ramírez Godina F, Ortega Ortiz H, Benavides Mendoza A, Robledo TV, Cabrera De la Fuente M. Use of chitosan-PVA hydrogels with copper nanoparticles to improve the growth of grafted watermelon. Molecules. 2017;**22**:1031. DOI: 10.3390/molecules22071031

[111] Amin AA, Rashad EL-SM, EL-Abagy HMH. Physiological effect of indole-3-butyric acid and salicylicacid on growth, yield and chemical constituents of onion plants. Journal of Applied Sciences Research. 2007;**3**(11):1554-1563

[112] Dzung NA, Thang NTE. Effect of oligoglucosamine prepared by enzyme degradation on the growth of soybean. Advances in Chitin and Chitosan Science. 2002;**5**:463-467

[113] Rahman M, Mukta JA, Sabir AA, Gupta DR, Mohi-Ud-Din M, Hasanuzzaman M, et al. Chitosan biopolymer promotes yield and stimulates accumulation of antioxidants in strawberry fruit. PLoS One. 2018;**13**(3):e0203769. DOI: 10.1371/journal.pone.0203769

[114] Hadwiger LA. Multiple effects of chitosan on plant systems: Solid science or hype. Plant Science. 2013;**208**:42-49. DOI: 10.1016/j.plantsci.2013. 03.007

[115] Lin W, Hu X, Zhang W, Rogers WJ, Cai W. Hydrogen peroxide mediates defence responses induced by chitosans of different molecular weights in rice. Journal of Plant Physiology. 2005;**162**:937-944. DOI: 10.1016/j.jplph.2004.10.003

[116] Hoat TX, Nakayashiki H, Yang Q, Tosa Y, Mayama S. Molecularcloning of the apoptosis-related calcium-binding protein AsALG-2in Avena sativa. Molecular Plant Pathology. 2013;**14**(3):222-229. DOI: 10.1111/j.1364-3703.2012.00844.x

[117] Ma ZX, Yang LY, Yan HX, Kennedy JF, Meng XH. Chitosan and oligochitosan enhance the resistance of peach fruit to brown rot. Carbohydrate Polymers. 2013;**94**(1):272-277. DOI: 10.1016/j.carbpol.2013.01.012

[118] Wang SY, Gao H. Effect of chitosan-based edible coating on antioxidants, antioxidant enzyme system, and postharvest fruit quality of strawberries (*Fragaria* x *aranassa* Duch.). LWT-Food Science and Technology. 2013;**52**(2):71-79. DOI: 10.1016/j.lwt.2012.05.003

[119] Siddaiah CN, Prasanth KV, Satyanarayana NR, Mudili V, Gupta VK, Kalagatur NK, et al. Chitosan nanoparticles having higher degree of acetylation induce resistance against pearl millet downy mildew through nitric oxide generation. Scientific

Reports. 2018;**8**(1):1-4. DOI: 10.1038/s41598-017-19016-z

[120] Borsani O, Valpuesta V, Botella MA. Evidence for a role of salicylic acid in the oxidative damage generated by NaCl and osmotic stress in Arabidopsis seedlings. Plant Physiology. 2001;**126**(3):1024-1030. DOI: 10.1104/pp.126.3.1024

[121] Karthikeyan M, Jayakumar V, Radhika K, Bhaskaran R, Velazhahan R, Alice D. Induction of resistance in host against the infection of leaf blight pathogen (*Alternaria palandui*) in onion (*Allium cepa* var aggregatum). Indian Journal of Biochemistry & Biophysics. 2005;**42**(6):371-377

[122] Yang TP, Poovaiah BW. Hydrogen peroxide homeostasis: Activation of plant catalase by calcium/calmodulin. The Proceedings of the National Academy of Sciences. 2002;**99**(6):4097-4102. DOI: 10.1073/pnas.052564899

[123] Xu QJ, Nian YG, Jin XC, Yan CZ, Jin LI, Jiang GM. Effects of chitosan on growth of an aquatic plant (*Hydrilla verticillata*) in polluted waters with different chemical oxygen demands. Journal of Environmental Sciences. 2007;**19**(2):217-221. DOI: 10.1016/S1001-0742(07)60035-7

[124] MacDonald MJ, D'Cunha GB. A modern view of phenylalanine ammonia lyase. Biochemistry and Cell Biology. 2007;**85**:273-282. DOI: 10.1139/o07-018

[125] Xing K, Zhu X, Peng X, Qin S. Chitosan antimicrobial and eliciting properties for pest control in agriculture: A review. Agronomy for Sustainable Development. 2015;**35**(2):569-588. DOI: 10.1007/s13593-014-0252-3

[126] Luna E, Pastor V, Robert J, Flors V, Mauch-Mani B, Ton J. Callose deposition: A multifaceted plant defense response. Molecular Plant-Microbe Interactions. 2011;**24**(2):183-193. DOI: 10.1094/mpmi-07-10-0149

[127] Mejía-Teniente L, de Dalia Durán-Flores F, Chapa-Oliver AM, Torres-Pacheco I, Cruz-Hernández A, González-Chavira MM, et al. Oxidative and molecular responses in *Capsicum annuum* L. after hydrogen peroxide, salicylic acid and chitosan foliar applications. International Journal of Molecular Sciences. 2013;**14**(5):10178-10196. DOI: 10.3390/ijms140510178

[128] Blume B, Nürnberger T, Nass N, Scheel D. Receptor-mediated increase in cytoplasmic free calcium required for activation of pathogen defense in parsley. The Plant Cell. 2000;**12**(8):1425-1440. DOI: 10.1105/tpc.12.8.1425

[129] Klüsener B, Young JJ, Murata Y, Allen GJ, Mori IC, Hugouvieux V, et al. Convergence of calcium signaling pathways of pathogenic elicitors and abscisic acid in Arabidopsis guard cells. Plant Physiology. 2002;**130**(4):2152-2163. DOI: 10.1104/pp.012187

[130] Tocci N, Simonetti G, D'Auria FD, Panella S, Palamara AT, Valletta A, et al. Root cultures of Hypericum perforatum subsp. angustifolium elicited with chitosan and production of xanthone-rich extracts with antifungal activity. Applied Microbiology and Biotechnology. 2011;**91**:977-987. DOI: 10.1007/s00253-011-3303-6

[131] Halim VA, Vess A, Scheel D, Rosahl S. The role of salicylic acid and jasmonic acid in pathogen defence. Plant Biology. 2006;**8**(03):307-313. DOI: 10.1055/s-2006-924025

[132] Rakwal R, Tamogami S, Agrawal GK, Iwahashi H. Octadecanoid signaling component "burst" in rice (*Oryza sativa* L.) seedling leaves upon

wounding by cut and treatment with fungal elicitor chitosan. Biochemical and Biophysical Research Communications. 2002;**295**:1041-1045. DOI: 10.1016/S0006-291X(02)00779-9

[133] Seo M, Koshiba T. Complex regulation of ABA biosynthesis in plants. Trends in Plant Science. 2002;**7**:41-48. DOI: 10.1016/S1360-1385(01) 02187-2

[134] Gutjahr C, Parniske M, Schekman R. Cell and developmental biology of arbuscular mycorrhiza symbiosis. Annual Review of Cell and Developmental Biology. 2013;**29**:593-617. DOI: 10.1146/annurev-cellbio-101512-122413

[135] Li K, Xing R, Liu S, Li P. Chitin and chitosan fragments responsible for plant elicitor and growth stimulator. Journal of Agricultural and Food Chemistry. 2020;*68*(44):12203-12211

[136] Liang Y, Tóth K, Cao Y, Tanaka K, Espinoza C, Stacey G. Lipochitooligosaccharide recognition: An ancient story. The New Phytologist. 2014;**204**:289-296. DOI: 10.1111/nph.12898

[137] O'Herlihy EA, Duffy EM, Cassells AC. The effects of arbuscular mycorrhizal fungi and chitosan sprays on yield and late blight resistance in potato crops from microplants. Folia Geobotanica. 2003;**38**:201-207. DOI: 10.1007/BF02803152

[138] Savci S. Investigation of effect of chemical fertilizers on environment. APCBEE Procedia. 2012;**1**(1):287-292. DOI: 10.1016/j.apcbee.2012.03.047

[139] Hussain MR, Devi RR, Maji TK. Controlled release of urea from chitosan microspheres prepared by emulsification and cross-linking method. Iranian Polymer Journal. 2012;**21**:473-479. DOI: 10.1007/s13726-012-0051-0

[140] Quesada HB, Baptista ATA, Cusioli LF, Seibert D, Bezerra CO, Bergamasco R. Surface water pollution by pharmaceuticals and an alternative of removal by low-cost adsorbents: A review. Chemosphere. 2019;**222**:766-780. DOI: 10.1016/j.chemosphere.2019.02.009

[141] Zhao L, Guan X, Yu B, Ding N, Liu X, Ma Q, et al. Carboxylated graphene oxide-chitosan spheres immobilize Cu^{2+} in soil and reduce its bioaccumulation in wheat plants. Environment International. 2019;**133**:105208. DOI: 10.1016/j.envint.2019.105208

[142] Qian J, Zhou L, Yang X, Hua D, Wu N. Prussian blue analogue functionalized magnetic microgels with ionized chitosan for the cleaning of cesium-contaminated clay. Journal of Hazardous Materials. 2020;**386**:121965. DOI: 10.1016/j.jhazmat.2019.121965

[143] Türker OC, Baran TA. Combination method based on chitosan adsorption and duckweed (*Lemna gibba* L.) phytoremediation for boron (B) removal from drinking water. International Journal of Phytoremediation. 2018;**20**(2):175-183. DOI: 10.1080/15226514.2017.1350137

[144] Sabeen AH, Kamaruddin SN, Noor ZZ. Environmental impacts assessment of industrial wastewater treatment system using electroless nickel plating and life cycle assessment approaches. International Journal of Environmental Science and Technology. 2019;**16**(7):3171-3182. DOI: 10.1007/s13762-018-1974-6

[145] Pérez Mora BE, Bellú SE, Mangiameli MF, García SI, González JC. Optimization of continuous arsenic biosorption present in natural contaminated groundwater. Journal of Chemical Technology & Biotechnology. 2019;**94**(2):547-555. DOI: 10.1002/jctb.5801

[146] Taha AA, Hameed NJ, Rashid FH. Preparation and characterization of (hyacinth plant/chitosan) composite as a heavy metal removal. Baghdad Science Journal. 2019;**16**(4):865-870. DOI: 10.21123/bsj.2019.16.4.0865

[147] Xiang J, Lin Q, Yao X, Yin G. Removal of Cd from aqueous solution by chitosan coated MgO-biochar and its in-situ remediation of Cd-contaminated soil. Environmental Research. 2021;**1**(195):110650. DOI: 10.1016/j.envres.2020.110650

[148] Dinh VP, Nguyen MD, Nguyen QH, Do TTT, Luu TT, Luu AT, et al. Chitosan-MnO_2 nanocomposite for effective removal of Cr (VI) from aqueous solution. Chemosphere. 2020;**257**:127-147. DOI: 10.1016/j.chemosphere.2020.127147

[149] Hu D, Lian Z, Xian H, Jiang R, Wang N, Weng Y, et al. Adsorption of Pb (II) from aqueous solution by polyacrylic acid grafted magnetic chitosan nanocomposite. International Journal of Biological Macromolecules. 2020;*154*:1537-1547. DOI: 10.1016/j.ijbiomac.2019.11.038 Epub November 13, 2019

[150] Lobo C, Castellari J, Lerner JC, Bertola N, Zaritzky N. Functional iron chitosan microspheres synthesized by ionotropic gelation for the removal of arsenic (V) from water. International Journal of Biological Macromolecules. 2020;*164*:1575-1583. DOI: 10.1016/j.ijbiomac.2020.07.253

[151] Zong H, Li K, Liu S, Song L, Xing R, Chen X, et al. Improvement in cadmium tolerance of edible rape (*Brassica rapa* L.) with exogenous application of chitooligosaccharide. Chemosphere. 2017;**1**(181):92-100. DOI: 10.1016/j.chemosphere.2017.04.024

[152] Rychter P. Chitosan/glyphosate formulation as a potential, environmental friendly herbicide with prolonged activity. Journal of Environmental Science and Health, Part B: Pesticides, Food Contaminants, and Agricultural Wastes. 2019;**54**(8):681-692. DOI: 10.1080/03601234.2019.1632644

[153] Rashidipour M, Maleki A, Kordi S, Birjandi M, Pajouhi N, Mohammadi E, et al. Pectin/chitosan/tripolyphosphate nanoparticles: Efficient carriers for reducing soil sorption, cytotoxicity, and mutagenicity of paraquat and enhancing its herbicide activity. Journal of Agricultural and Food Chemistry. 2019;**67**:5736-5745. DOI: 10.1021/acs.jafc.9b01106

[154] Yuan X, Wu Z, Li H, Wang Y, Liu F, Cai H, et al. Biochemical and proteomic analysis of 'Kyoho' grape (*Vitis labruscana*) berries during cold storage. Postharvest Biology and Technology. 2014;**88**:79-87. DOI: 10.1016/j.postharvbio.2013.10.001

[155] Li M, Chen C, Xia X, Garba B, Shang L, Wang Y. Proteomic analysis of the inhibitory effect of chitosan on *Penicillium expansum*. Food Science & Technology. 2019;**40**:250-257. DOI: 10.1590/fst.40418

Chapter 6

Chitinase from Basal Trypanosomatids and Its Relation to Marine Environment: New Insights on *Leishmania* Genus Evolutionary Theories

Felipe Trovalim Jordão, Aline Diniz Cabral, Felipe Baena Garcia, Edmar Silva Santos, Rodrigo Buzinaro Suzuki, Max Mario Fuhlendorf and Márcia Aparecida Sperança

Abstract

Leishmaniasis, an infectious disease that affects humans, domestic dogs, and wild animals, is caused by 20 of the 53 Leishmania genus species and is transmitted by sandflies. Despite its significant impact, the disease is often neglected. *Leishmania* genus, belong to Trypanosomatide Family and Kinetoplastida Order, are grouped in five subgroups according to biogeographic and evolution history of parasites and hosts. The GH18 Leishmania chitinase is encoded by a specie-specific single copy gene, conserved in basal groups of trypanosomatids, and is absent in the genus *Trypanosoma*. Preservation of the chitinase genomic locus in the aquatic free-living protozoan Bodo saltans, discloses a primitive common origin. Trypanosomatid chitinase amino acid sequence comparative analysis revealed high similarity with chitinase from sea living prokaryotes and protozoan microorganisms, indicating a probable marine origin. Amino acid sequence comparative analysis revealed that perhaps the trypanosomatid chitinase derived from a water living Kinetoplastida ancestor and its phylogenetic reconstruction corroborates the Supercontinent Origins theory for *Leishmania*. The chitinase-encoding gene was effective for differential molecular diagnosis among *Leishmania* clinical important species worldwide.

Keywords: leishmaniasis, molecular diagnosis, molecular evolution, chitinase, marine environment

1. Introduction

Leishmania genus protozoan parasites are the causative agent of leishmaniases, a complex of diseases that affect the tegument or the viscera. Leishmania parasites are transmitted among humans, domestic dogs, and wild animal hosts by insect vectors of the Psycodidae Family (sandflies) as well as the *Phlebotomus* (Old World) and *Lutzomyia* genus (New World - the Americas) [1]. Up to now 53 *Leishmania* species were described which are divided into five groups, the subgenera *Leishmania, Viannia, Sauroleishmania, Mundini,* and *Paraleishmania*. Most of the *Leishmania* species are zoonotic and 20 are incriminated to cause disease in human [1].

The diseases caused by *Leishmania* parasites, named leishmaniases, can be divided into Tegumentar (TL) and Visceral (VL), depending on the species of infecting parasite and host immunity conditions. Leishmaniasis can range from mild tegumentar ulcerations to fatal visceral infection. Leishmania parasites are endemic in 98 countries distributed in all continents and its prevalence is estimated as 0.4 and 1.2 million cases of VL and TL, respectively [2]. In the Americas, Brazil accounts for the highest incidence of leishmaniasis with wide spreading of TL and VL in expansion [3].

The VL is in second place as to the highest impact on health population, just behind malaria, in India and in the Mediterranean countries being caused by *L. (Leishmania) donovani* and *L. (Leishmania) infantum*, respectively [4]. In the Americas VL is caused by *L. infantum*, where this species of parasite is not endemic and probably it entered the Americas by infected dogs brought by Mediterranean colonizers [5]. Seventy-five percent of TL new cases occur in Afghanistan, Algeria, Colombia, Brazil, Iran, Syria, Ethiopia, Sudan, Costa Rica, and Peru, causing morbidity and disfiguration in infected people [6]. In South America, TL is mainly caused by the most prevalent endemic *Leishmania* species, viz., *L. (Viannia) braziliensis*, *L. (Leishmania) amazonensis,* and *L. (Leishmania) mexicana*. In spite of much efforts, a precise diagnostic test and effective treatment for leishmaniasis are still unavailable [7]. Thus, a detailed understanding of all aspects of specific biology and host-parasite relationships are important prior to facilitating the formulation of innovative and effective drugs and diagnostic tests for developing adequate prevention and control strategies.

The Leishmania parasites to complete their life cycle, they must invade the digestive tract of sand-fly species, which requires the degradation of insect chitin by a specific parasite chitinase [8]. Chitinase catalyzes the β-1,4-glycoside bond hydrolysis reaction of N-acetylglucosamine of chitin and chitodextrins [9]. Amino acid sequence similarity analysis of Leishmania GH18 chitinase grouped these enzymes in the GH18 and GH19 glycosyl hydrolase families. Initial studies in *Leishmania* revealed that chitinase and N-acetylglucosaminidase activities were found in promastigote supernatant cultures of *L. (Leishmania) major*. Similar activity of both enzymes was observed in *L. donovani, L. infantum, L. braziliensis, Leptomonas seymouri, Crithidia fasciculate,* and *Trypanosoma lewisi*. The chitinolytic action was attributed to the parasite secretion and was not secreted through the sand fly gut [10, 11]. The gene encoding a GH18 chitinase was initially obtained for *L. donovani* (Ld Cht1) using molecular approach and biochemical characterization. Molecular genetic studies enabled the identification of a similar gene in several species of *Leishmania* genus (*L. major, L. infantum, L. donovani,* and *L. braziliensis*) [12].

The biological importance of the Leishmania GH18 chitinase in parasite life cycle was confirmed after homologous episomal overexpression of chitinase in both amastigotes and promastigotes of *L. Mexicana*. In the insect vector, overexpression of Leishmania chitinase resulted in an increase in transmission rate where in the vertebrate host ensued increased pathogenicity, thereby indicating that chitinase plays an important role in parasite development, survival, and transmission in mammalian hosts [13, 14]. Due to the biological properties and characteristics of *Leishmania* genus GH18 chitinase, including *locus* conservation, species-specific amino acid and nucleotide sequence, expression in all species of parasite developmental stages, and extracellular exportation, this investigation focused on the potential of the chitinase-encoding gene as a molecular diagnostic tool and a phylogenetic marker for studying basal trypanosomatids groups.

2. Chitinase genetic *locus* conservation among Kinetoplastida revealed its marine origin

Evaluation of the phylogenetic relationship of the GH18 family chitinase in Kinetoplastida was performed through comparative analysis by Basic Local Alignment Search Tool (BLAST) on chitinase amino acid and nucleotide sequences of trypanosomatids, available in public databanks (**Figure 1**) [15]. The results showed that the chitinase amino acid sequence is highly conserved among the species of the *Leishmania* genus, with identity variations ranging from 78 to 100%. Additionally, basal trypanosomatids such as *Leptomonas*, *Strigomonas,* and *Angomonas* also harbor a GH18 chitinase that is similar to that of *Leishmania,* with identity percentages of 60, 40, and 35%, respectively. This chitinase was found to be absent in parasites from *Trypanosoma* genus. A protein with identity of 32% with *Leishmania* genus chitinase was also present in the free-living acquatic protozoa from Kinetoplastida Order and Bodonidae Family, *Bodo saltans*, which is used as external group of Trypanosomatid in phylogenetic studies. These results suggest the occurrence of a homologous GH18 chitinase in a Kinetoplastida ancestor.

The TritrypDB genomic resource tools were utilized to analyze the locus of the chitinase-encoding gene in all available Kinetoplastida sequences, comprising *Bodo saltans*, *Leptomonas, Angomonas, Strigomonas,* and *Leishmania*. The results indicated a high degree of conservation, which further supports the hypothesis of a shared origin. In all organisms included in the analysis, the GH18 chitinase is a single copy gene (**Figure 1**).

BLAST analysis of the chitinase amino acid sequence from *Bodo saltans* showed identity of 38% with the chitinase of the marine microorganisms *Perkinsus marinus* and *Micromonas pusilla*, indicating that the GH18 chitinase of the trypanosomatids ancestor emerged from marine environment. These data are supported by the phylogenetic reconstruction of Kinetoplastida GH18 chitinase which grouped in a common cluster, separated from the GH18 chitinases of human and insects (**Figure 2**).

3. The phylogenetic relationship among basal trypanosomatids chitinase corroborated the supercontinent origin hypothesis for *Leishmania* genus

The barcode for phylogenic relationship among trypanosomatids corresponds to the variable region V4 (described also as V7 and V8) of the small ribosomal subunit (V4 rRNA SSU) [5, 16]. In order to investigate the potential of chitinase-encoding gene for trypanosomatids phylogenetic reconstruction, a partial 953 bp chitinase-encoding gene fragment and the corresponding amino acid sequence from trypanosomatids available in genomic databanks and generated by our research group was evaluated for phylogenetic reconstruction by *Neighbor-Joining* [17] method, using the MEGA4 software (**Figure 3**) (**Table 1**). The obtained trees were compared to the phylogenetic analysis performed by *Neighbor-Joining* method, with the trypanosomatid barcode of the same species (**Figure 4**). The phylogenetic reconstruction of *Leishmania* based on GH18 chitinase-encoding gene corroborated the thesis of the

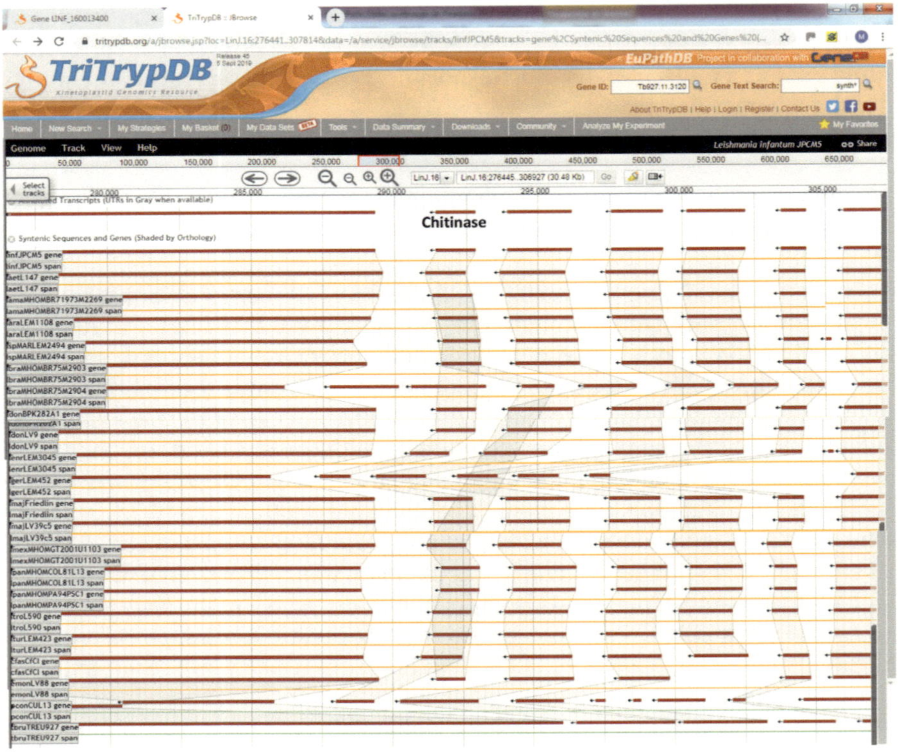

Figure 1.
Leishmania species CH18 chitinase locus obtained from available sequences in the Tritrypdb data base, a trypanosomatide genomic bank, using the software for genomic analysis (https://tritrypdb.org/a/jbrowse.jsp?loc=LinJ.16:276441..307814&data=/a/service/jbrowse/tracks/linfJPCM5&tracks=gene%2CSyntenic%20Sequences%20and%20Genes%20(Shaded%20by%20Orthology).

Supercontinent Origin for *Leishmania* genus [18] with higher accuracy, according to bootstrap results, when compared to the trypanosomatid barcode. The rRNA SSU V4 groups the subgenera *Mundinia* together with the subgenera *Viannia* (**Figure 4**), while the phylogenetic relationships based on the chitinase protein separate the species of all *Leishmania* subgenera and *Paraleishmania* group.

4. Conventional polymerase chain reaction (PCR) associated to restriction length polymorphism differentiated *Leishmania* subgenera of old and New World

In silico analysis of the Genbank *Leishmania* species chitinase, which belongs to the glycosil-hydrolase 18 (GH18) family, revealed that it is localized on chromosome 16 and encoded by a single copy gene. The analysis also indicated high inter-subgenera identity in all crucial putative domains and post-translational modifications [21, 22].

Detection and identification of *Leishmania* parasite subgenera was successfully obtained by employment of P*st I* restriction analysis on the *Leishmania* species chitinase gene 953 bp fragment. Using D*de I* restriction analysis was possible to separate viscerotropic and tegumentar species from Old World *Leishmania* subgenus and *Sauroleishmania* and *Viannia* subgenera (**Table 2**).

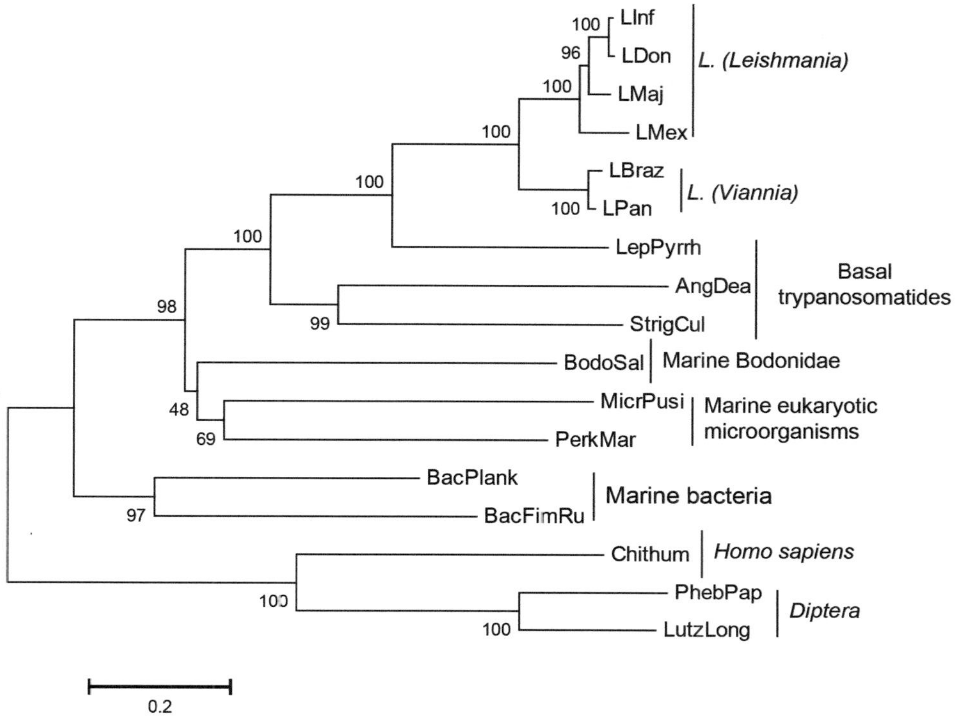

Figure 2.
Evolutionary relationships of GH18 chitinase amino acid complete sequence by neighbor-joining method. Phylogenetic reconstruction was conducted in MEGA4 and the figure corresponds to the optimal tree with the sum of branch length = 6.42769400. The bootstrap values (1000 replicates) are shown next to the branches. Evolutionary distances used to infer the phylogenetic tree are drawn to scale.

5. Discussion

Several regions, such as Brazil, are endemic to more than one species of Leishmania, with both VL (*L. infantum*) and TL (*L. braziliensis, L. amazonensis*). Additionally, there are parasites from the *Trypanosoma* genera (*T. cruzi, T. rangeli*) that show cross-reaction in various serological and molecular tests. Thus, the differential molecular diagnosis of *Leishmania* based on chitinase-encoding gene is highly sensitive and specific and becomes relevant in these multiple species of *Leishmania* and Trypanosoma endemic areas. Therefore, a diagnostic method capable of distinguishing between different *Leishmania* species in animal, human, and vector reservoirs will better guide leishmaniasis control.

Nucleic acid detection techniques in samples from people and/or animals infected with *Leishmania*, such as PCR, are used for detection and identification of the parasite since the 1980s. PCR includes the amplification of various fragments, including those of the gene that encodes the small ribosomal RNA subunit (SSU rDNA) [23], the transcribed internal ribosomal DNA spacer (ITS) [24], sequences that correspond to kinetoplast (kDNA) [25], and mini-exon [26], as well as the gene encoding the heat shock protein HSP70 [27], etc. Despite its high sensitivity and, depending on the molecular target, high specificity, PCR is more commonly utilized in epidemiological studies rather than as a routine diagnostic method. In addition, to achieve

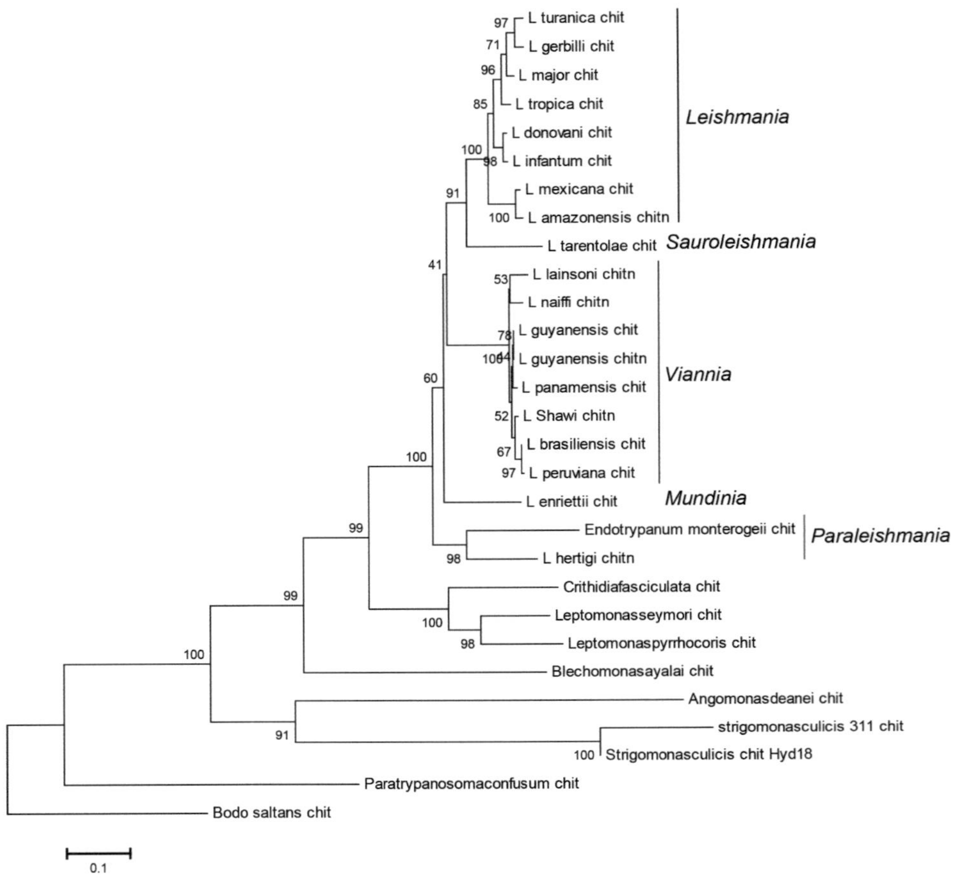

Figure 3.
Evolutionary relationships of Leishmania genus and ancient trypanosomatides by neighbor-joining method using the 292 amino acid residues corresponding to the 949 bp chitinase-encoding gene fragment. Phylogenetic reconstruction was conducted in MEGA4 and the figure corresponds to the optimal tree with the sum of branch length = 4.73393897. The bootstrap values (1000 replicates) are shown next to the branches. Evolutionary distances used to infer the phylogenetic tree are drawn to scale. Leishmania subgenus and Paraleishmania parasites are described at right.

high sensitivity in the methodologies evaluated so far, PCR complementation with other techniques including nested PCR and hybridization is required. To identify *Leishmania* species, various methodologies are employed, including the analysis of restriction fragment sizes of PCR products. However, as most gene targets have multiple copies, interpretation of the results can increase the difficulty of using these techniques in clinical routine. In addition, false positives are possible due to contamination with other post-PCR amplified samples or DNA fragments and cross-reaction with other pathogens, including *Trypanosoma* [28].

The differential diagnosis of *Leishmania* subgenera based on chitinase-encoding gene offers certain advantages over other molecular methods. This is because it is encoded by a single copy gene, which is absent in the *Trypanosoma* genus, thereby allowing for specific detection of *Leishmania* parasites. Also, the sensitivity of the method, regarding the size of the amplified fragment, is high, supporting post-PCR analysis of a single reaction obtained directly from

Species	Gene	Source	Observations
Angomonas deanei	Chitinase	ENA[1]:EPY25377	
A. deanei	SSU	Genbank[2]:HM593012	Isolate TCC080E
Blechomonas ayalai	Chitinase	Tritrypdb[3]:rna_Baya_138_0020-1	B08-376
B. ayalai	SSU	Genbank KF054116	Isolate B08-376
Bodo saltans	Chitinase	Genbank CYKH01001162	Sequência do *locus*
B. saltans	SSU	Genbank MF962814	Strain NG
Crithidia luciliae	SSU	Genbank KY364901	Isolate ATCC14765
Critidia fasciculate	Chitinase	Tritrypdb::CFAC1_120016000	Strain Cf-Cl
Endotrypanum monterogeii	Chitinase	Tritrypdb:EMOLV88_160012400.1	Strain LV88
E. monterogeii	SSU	Genbank X53911	
Fimbriiglobus ruber	Chitinase	Genbank OWK46432.1	Bacterium from Order *Planctomycetales*, associated to Crustacea.
Homo sapiens	Chitinase	Genbank AAG10644.1	Chitotriosidase, Macrofago, CH1; GH18.
L. amazonensis	Chitinase	Genbank: MG869127	Strain IOCL 0575
L. amazonensis.	SSU	Genbank JX030083	Isolate LaPH8
L. braziliensis	Chitinase	Genbank LS997615	MHOM/BR/75/M2904
L. braziliensis	SSU	Genbank JX030135	Isolate LbrET
L. donovani	Chitinase	Genbank CP019523	Strain MHOM/IN/1983/AG83
L. donovani	SSU	Genbank X07773	
L. enrietti	Chitinase	Tritrypdb:LEN LEM3045_160013500.1	Strain LEM 3045
L. enrietti	SSU	Genbank KF041798	Isolate CBT47
L. gerbilli	Chitinase	Tritrypdb:LGELEM452_160013100.1	*Strain LEM452*
L. guyanensis	Chitinase	ENA:CCM15041	
L. guyanensis	SSU	Genbank X53913	
L. gymnodactyli	SSU	Genbank KX790780	TCC722
L. hoogstraali	SSU	Genbank KF041810	Isolate L7
L. infantum	Chitinase	Genbank:FR796448	Strain JPCM5
L. infantum chagasi	SSU	Genbank KT240047	Isolate CBT 153
L. lainsoni	SSU	Genbank KF041805	Isolate IOCL 1023
L. major	Chitinase	Genbank:FR796412.1	Strain Friedlin
L. major	SSU	Genbank XR_002460813	
L. Mexicana	Chitinase	Genbank:AY572789	
L. Mexicana	SSU	Genbank GQ332360	Strain Friedlin
L. naiffi	SSU	Genbank KF041807	Isolate IOCL 1365

Species	Gene	Source	Observations
L. panamensis	Chitinase	Genbank CP009385	Strain MHOM/PA/94/PSC-1
L. panamensis	SSU	Genbank JN003595	Strain UA946
L. peruviana	Chitinase	Genbank LN609244	PAB 4377
L. shawi	SSU	Genbank KF041808	Isolate IOCL 1545
L. tarantolae	Chitinase	Tritrypdb:LtaP16.0770	Parrot-TarII
L. tarentolae	SSU	Genbank X53916	
L. tropica	Chitinase	TritrypdbLTRL590_160013800.1	Strain L590
L. turanica	Chitinase	Tritrypdb:LTULEM423_160013000.1	Strain LEM 423
Leishmania sp	SSU	Genbank AF303938	MHOM/MQ/92/MAR1
Leptomonas pyrrhocoris	Chitinase	Tritrypdb:rna_LpyrH10_15_0870	H10
L. pyrrhocoris	SSU	Genbank JN036653	Isolate H10
Leptomonas seymori	Chitinase	Tritrypdb:PCLsey_0068_0030	ATCC 30220
L. seymouri	SSU	Genbank KP717894	Isolate Ld_39
Lutzomyia longipalpis	Chitinase	Genbank AAN71763.1	GH18 chitinase
M. pusilla	Chitinase	Genbank XP_003063458.1	Marine photosynthetic eukaryotic microorganism.
Paratrypanosoma confusum	Chitinase	Tritrypdb:PCON_0062580	Isolate cul13
P. confusum	SSU	Genbank KF963538	Isolate cul13
P. marinus	Chitinase	Genbank XM_002788039.1	Eukaryotic microorganism, pathogen of oysters.
Phlebotomus papatasi	Chitinase	Genbank AAV49322.1	GH18 chitinase
Planktomyces sp	Chitinase	Genbank OAI56776.1	Bacterium from marine environment.
Sergeia podlipaevi	SSU	Genbank DQ394362	*Strain* CER3
Strigomonas culicis	Chitinase	ENA:EPY22137	522 amino acids
S. culicis	Chitinase	ENA:EPY29957	311 amino acids
S. culicis	SSU	Genbank HM593009	Isolate TCC012E
Trypanosomatidae sp	SSU	Genbank AF071866	Strain EVA

[1]*ENA: European Nucleotide Archive (Available from: http://www.ebi.ac.uk/ena/data/view/<accession>).*
[2]*Genbank [19].*
[3]*TritrypDB [20]: Kinetoplastid Genomics Resource.*

Table 1.
Nucleotide sequences information.

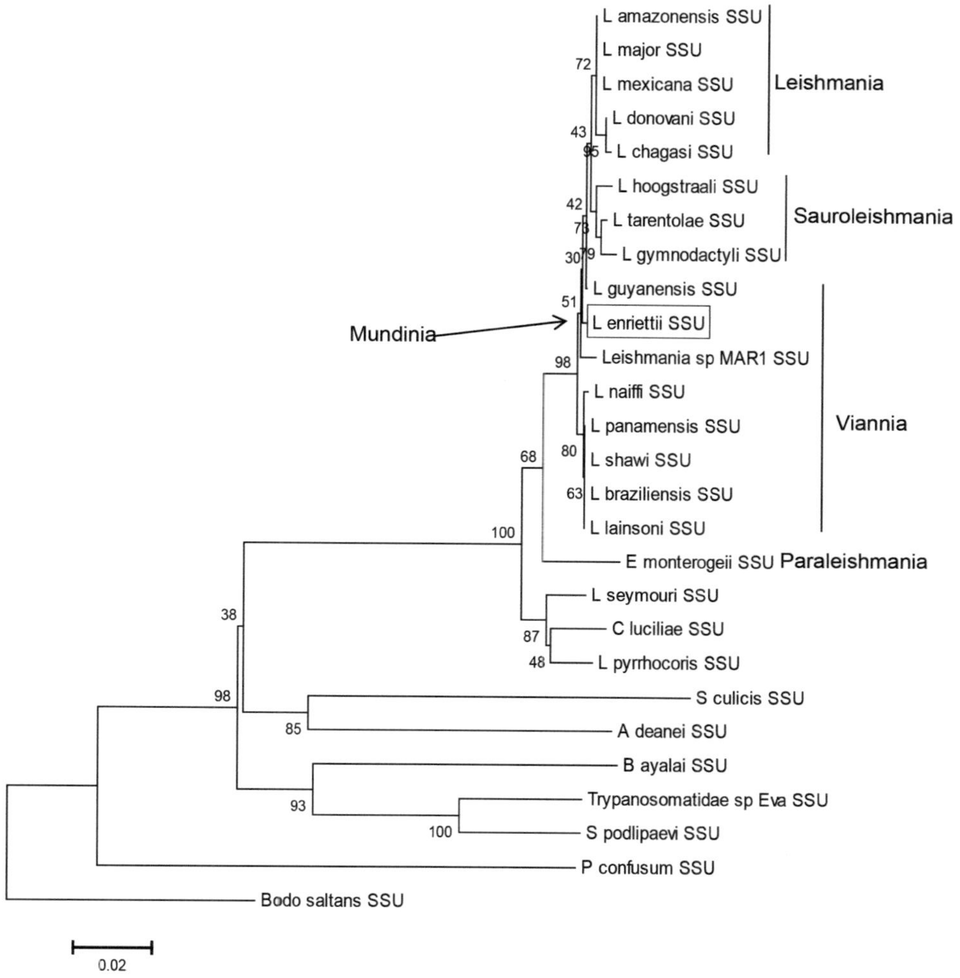

Figure 4.
Evolutionary relationships of Leishmania genus and ancient trypanosomatides by neighbor-joining method using the 834 nucleotide residues corresponding to the small RNA subunit variable region 4 (V4 rRNASSU) encoding gene. Phylogenetic reconstruction was conducted in MEGA4 and the figure corresponds to the optimal tree with the sum of branch length = 0.80513269. The bootstrap values (1000 replicates) are shown next to the branches. Evolutionary distances used to infer the phylogenetic tree are drawn to scale. Leishmania subgenus and Paraleishmania parasites are described at right. Box: L.enrietti classified in subgenus Mundinia.

biological samples. Restriction analysis of the 953 bp *Leishmania* chitinase PCR fragment with P*stI* permitted the identification of medical important species in Latin America where three different *Leishmania* subgenera circulate in animal reservoirs, human, and sandflies [21]. Given the specificity of the *Leishmania* chitinase-encoding gene, the molecular diagnostic method can also be used to identify isolated parasites from biological samples, with high specificity, by restriction analysis and/or sequencing [29]. Also, using the restriction enzyme D*de I* on the 953 bp chitinase PCR fragment, it is possible to differentiate *L. major* from all others Old World *Leishmania* subgenus species, which is of clinical importance in Oriental TL endemic countries (**Table 2**) [30].

Subgenera	Species	Geographic distribution	*Dde* I fragments (bp)	*Pst I* fragments (bp)
Leishmania	*L.donovani*	Old World	608, 157,188	390, 113, 258, 192
Leishmania	*L. infantum*	New and Old World	608, 157,188	390, 113, 258, 192
Leishmania	*L. turanica*	Old World	608, 157,188	390, 113, 258, 192
Leishmania	*L. gerbili*	Old World	608, 157,188	390, 113, 258, 192
Leishmania	*L. major*	Old World	58, 550, 79, 78, 188	390, 113, 258, 192
Leishmania	*L. mexicana*	New World	602, 6, 157, 188	390, 113, 450
Leishmania	*L. amazonensis*	New World	602, 6, 157, 188	390, 113, 450
Viannia	*L. braziliensis*	New World	325, 284, 344	503, 450
Sauroleishmania	*L. tarantolae*	Old World	765, 188	503, 450
Mundinia	*L. enrietti*	New World	112,292, 180, 25, 344	503, 258, 192

Table 2.
Restriction fragment sizes of the 953 bp chitinase PCR amplicon digested with Dde I and Pst I.

The origins of parasites from *Leishmania* genus and its evolutionary relationships are investigated through phylogenetic reconstructions associated to data on biogeographic dispersion and evolution of their vertebrates and sandflies corresponding hosts, being a matter of discussion on conflicting information [1]. The revision by Akhoundi, et al. 2016, details the three principal theories proposed for *Leishmania* origin, a Palearctic, Neotropical, and a Neotropical/African/Multiple Origins. The most widely accepted theory regarding the origin of *Leishmania* is the Supercontinent hypothesis, a variation of the Multiple Origins hypothesis. This theory suggests that the *Viannia* and *Leishmania* subgenera evolved independently during the separation of South America from Africa. The Supercontinent hypothesis proposes that *Leishmania* originated on Gondwana and evolved from monoxenous parasites [18]. This theory is supported by biogeographic data and animal host migration patterns, and was developed through phylogenetic reconstruction using a large multi-gene dataset (over 200,000 informative sites) [18]. In this work, the phylogenetic reconstruction of *Leishmania* genus with the 953 bp partial chitinase-encoding gene sequence corroborates the Supercontinent hypothesis, grouping species of each *Leishmania* subgenera and the Paraleishmania group (**Figure 4**).

The genomic locus of GH18 chitinase-encoding gene is conserved among basal trypanosomatids, including the *B. saltans*. Besides, amino acid sequence comparison studies among GH18 chitinases from trypanosomatids using public genome databanks revealed 35% of identity of GH18 chitinases of marine protozoa and bacteria to the similar enzyme of *B. saltans*. These results strongly suggest that the GH18 chitinase from Kinetoplastida group derived from a common marine ancestor, harboring the primitive enzyme.

The phylogenetic position of subgenus *Sauroleishmania,* according to the Supercontinent hypothesis, indicates the switch of its *Leishmania* ancestors from mammalian to reptilian hosts [1]. In considering a probable marine environment emergence of the trypanosomatid GH18 chitinase, a highly conserved and unique gene in basal groups, including *Leishmania*, it is possible to explore that the *Sauroleishmania* subgenus could diverge from an ancestor before the rise of mammals, during the transition of animals from marine to the terrestrial environment. In this case, *Leishmania*-like parasites could be found in fish and amphibians. Considering the conservation of the chitinase-encoding gene in *Leishmania* group, the diagnostic method developed in this work can be used to investigate this hypothesis directly on biological samples, circumventing the isolation difficulties of unknown *Leishmania*-like parasites.

The barcode gene used for phylogenetic studies in trypanosomatids corresponds to the V4 rRNA SSU. However, for basal groups of trypanosomatids, the chitinase-encoding gene and amino acid sequence presented better results, as demonstrated through comparison between phylogenetic trees generated by Neighbor-joining method for both markers (**Figures 3** and **4**). Furthermore, the 953 bp chitinase fragment can be obtained by PCR directly from biological samples, while the V4 rRNA fragment is obtained preferentially from isolated parasites. These results indicate that the partial sequence of the gene encoding trypanosomatids chitinase can be used as a barcode to investigate the phylogenetic relationships among basal species of the group.

In silico prediction of the biochemical and molecular characteristics of *Leishmania* chitinase, such as high solubility and exportation of the native protein to the extracellular medium, associated to species-specific sequences, as well as the absence of a similar gene in the genus *Trypanosoma*, indicated its potential use as an antigen in accurate differential serological diagnosis for *Leishmania* species. Thus, in order to produce high amounts of *Leishmania* chitinase from the four different species representing the three major taxonomic groups from the subgenera *Viannia* (*L. braziliensis*) and *Leishmania* (*L. amazonensis*/*L. mexicana* and *L. infantum*), a conventional prokaryotic expression system was initially used. However, after tests under various expression conditions, the proteins remained insoluble, probably due to unfolding, the absence of the predicted N-glycosylation signal, and the formation of inclusion bodies [21]. The homologous chitinase of *L. donovani* was partially soluble in *E. coli* only when expressed in fusion with thioredoxin [31], thus indicating that even when using different strategies, the prokaryotic system is not appropriate for producing a correct folded *Leishmania* chitinase, essential in obtaining an effective antigen for serological testing, for example.

After observing the successful expression of *L. major* GP63, a glycosylated membrane protein, in insect cells using recombinant baculovirus [32], we obtained several baculovirus constructions of chitinase from the four *Leishmania* species by utilizing the *Bac-to-Bac* insect cell expression system. However, despite several attempts, it was found that this was not the case. This suggests that there may be intrinsic molecular characteristics associated with the insolubility of *Leishmania* species chitinase, such as post-translational signals that are incompatible with the organelle machinery of insect cells. Consequently, this may lead to the accumulation of unfolded recombinant proteins in cellular compartments. Investigation by way of *in situ* detection of recombinant proteins from constructions with N and C terminal histidine-tags, using Fluorescein Isothiocyanate (FITC) labeled anti-histidine antibody, revealed their accumulation in the plasma membrane [21]. This was confirmed by Thin Layer Chromatography (TLC) of insect cell membranes infected with *L. infantum* chitinase baculoviruses previously enriched with recombinant chitinase, and presenting lipid chemical signatures of plasma membrane. Recombinant forms of *Leishmania* species chitinase accumulation in plasma membrane, independent of differential physicochemical characteristics, presuppose the occurrence of a specific molecular feature necessary for precise *Leishmania* chitinase folding, such as chaperones.

Chaperones play an essential role in the regulation of biological functions by facilitating changes in the conformation of non-native protein. This is achieved through several mechanisms, including ribosome nascent polypeptide folding, organelle and cellular protein-membrane addressing, and the disassembly of macromolecular aggregates. Biochemical modifications, such as phosphorylation networks and hydrophobic interactions, also contribute to this process [33, 34]. There are several eukaryotically conserved and divergent chaperone families in *Leishmania* parasites that could explain the need for specific chitinase-chaperone interaction in correct folding [35], such as cyclophilin, a specific *L. donovani* chaperone, involved in reversing ADP-dependent inactive aggregates of adenosine kinase under physiological conditions, and a key enzyme in the leishmanial purine salvage pathway [36]. Further evidence of specific chitinase-chaperone interaction comprises the highly conserved *Leishmania* genus chitinase C-terminal ATP-dependent kinase domain (Data in Brief co-publication) since biochemical regulation of several chaperone functions is often associated to ATP-dependent phosphorylation [37]. Further studies of native *Leishmania* chitinase are required to investigate this hypothesis.

Leishmania chitinase is specific to basal groups of trypanosomatids genera, probably derived from an ancestor living in a marine environment, and unique in the human pathogen group. There are no *Leishmania* chitinase or homologous proteins described with a molecular structure associated to biochemical function. Considering biological importance and *Leishmania* genus chitinase specificity, novel molecular-based studies of native protein should shed light on its biochemical function, thereby facilitating its use not only in diagnosis, but also in drug and vaccine designs for controlling and treating leishmaniasis.

6. Conclusions

- Phylogenetic analysis of kinetoplastida chitinase revealed an ancient origin;
- Chitinase phylogeny corroborates the Supercontinent Origins theory for *Leishmania*;
- Chitinase is conserved among basal trypanosomatids and absent in *Trypanosoma* genus;
- Diagnosis based on *Leishmania* chitinase differentiates Old and New World Subgenera;
- Amino acid comparison analysis supports a marine origin for *Leishmania* chitinase.

Acknowledgements

This work was funded by Fundação de Amparo a Pesquisa do Estado de São Paulo (FAPESP) (Grant numbers 2016/14514-4 and 2012/20221-9; fellowships 2013/26096-4 and 2018/05133-2).

Conflict of interest

The authors declare no conflict of interest.

Author details

Felipe Trovalim Jordão[1], Aline Diniz Cabral[2], Felipe Baena Garcia[1],
Edmar Silva Santos[1], Rodrigo Buzinaro Suzuki[3], Max Mario Fuhlendorf[1]
and Márcia Aparecida Sperança[1*]

1 Federal University of ABC, Center for Natural and Human Sciences, São Bernardo do Campo, São Paulo, Brazil

2 Department of Parasitology, Federal University of Uberlandia, Uberlãndia, Brazil

3 Marilia University, São Paulo, Brazil

*Address all correspondence to: marcia.speranca@ufabc.edu.br

References

[1] Akhoundi M, Kuhls K, Cannet A, Votypka J, Marty P, Delaunay P, et al. A historical overview of the classification, evolution, and dispersion of leishmania parasites and sandflies. PLoS Neglected Tropical Diseases. 2016;**10**(3):e0004349. Epub 20160303. DOI: 10.1371/journal.pntd.0004349

[2] Alvar J, Velez ID, Bern C, Herrero M, Desjeux P, Cano J, et al. Leishmaniasis worldwide and global estimates of its incidence. PLoS One. 2012;7(5):e35671. Epub 20120531. DOI: 10.1371/journal.pone.0035671

[3] Georgiadou SP, Makaritsis KP, Dalekos GN. Leishmaniasis revisited: Current aspects on epidemiology, diagnosis and treatment. Journal of Translational Internal Medicine. 2015;**3**(2):43-50. Epub 20150630. DOI: 10.1515/jtim-2015-0002

[4] WHO. Working to Overcome the Global Impact of Neglected Tropical Diseases: First WHO Report on Neglected Tropical Diseases. Geneva, Switzerland: World Health Organization; 2010

[5] Marcili A, Speranca MA, da Costa AP, Madeira Mde F, Soares HS, Sanches Cde O, et al. Phylogenetic relationships of Leishmania species based on trypanosomatid barcode (SSU rDNA) and gGAPDH genes: Taxonomic revision of Leishmania (L.) infantum chagasi in South America. Infection, Genetics and Evolution. 2014;**25**:44-51. Epub 20140416. DOI: 10.1016/j.meegid.2014.04.001

[6] WHO. Weekly epidemiological record. World Health Organization International. 2016;**91**(22):11

[7] Souza W. Doenças Negligenciadas. Academia Brasileira de Ciências. 2010:56. ISBN: 978-85-85761-30-1

[8] Shahabuddin M, Vinetz JM. Chitinases of human parasites and their implications as antiparasitic targets. EXS. 1999;**87**:223-234. DOI: 10.1007/978-3-0348-8757-1_16

[9] Cohen-Kupiec R, Chet I. The molecular biology of chitin digestion. Current Opinion in Biotechnology. 1998;**9**(3):270-277. Epub 1998/07/03

[10] Schlein Y, Jacobson RL, Shlomai J. Chitinase secreted by Leishmania functions in the sandfly vector. Proceedings of the Biological Sciences. 1991;**245**(1313):121-126. Epub 1991/08/22. DOI: 10.1098/rspb.1991.0097

[11] Rogers ME, Chance ML, Bates PA. The role of promastigote secretory gel in the origin and transmission of the infective stage of Leishmania mexicana by the sandfly Lutzomyia longipalpis. Parasitology. 2002;**124**(Pt 5):495-507 Epub 2002/06/07

[12] Shakarian AM, McGugan GC, Joshi MB, Stromberg M, Bowers L, Ganim C, et al. Identification, characterization, and expression of a unique secretory lipase from the human pathogen Leishmania donovani. Molecular and Cellular Biochemistry. 2010;**341**(1-2):17-31. Epub 2010/03/30. DOI: 10.1007/s11010-010-0433-6

[13] Joshi MB, Rogers ME, Shakarian AM, Yamage M, Al-Harthi SA, Bates PA, et al. Molecular characterization, expression, and in vivo analysis of LmexCht1: The chitinase of the human pathogen, Leishmania mexicana. The Journal of Biological Chemistry. 2005;**280**(5):3847-3861. Epub 2004/11/25. DOI: 10.1074/jbc.M412299200

[14] Rogers ME, Hajmova M, Joshi MB, Sadlova J, Dwyer DM, Volf P,

et al. Leishmania chitinase facilitates colonization of sand fly vectors and enhances transmission to mice. Cellular Microbiology. 2008;**10**(6):1363-1372. Epub 2008/02/21. DOI: 10.1111/j.1462-5822.2008.01132.x

[15] O'Driscoll A, Belogrudov V, Carroll J, Kropp K, Walsh P, Ghazal P, et al. HBLAST: Parallelised sequence similarity–A hadoop mapreducable basic local alignment search tool. Journal of Biomedical Informatics. 2015;**54**:58-64. Epub 2015/01/28. DOI: 10.1016/j.jbi.2015.01.008

[16] Rackevei AS, Bogers A, Engstler M, Dandekar T, Wolf M. About the analysis of 18S rDNA sequence data from trypanosomes in barcoding and phylogenetics: Tracing a continuation error occurring in the literature. Biology. 2022;**11**:7. DOI: 10.3390/biology11111612

[17] Saitou N, Nei M. The neighbor-joining method: A new method for reconstructing phylogenetic trees. Molecular Biology and Evolution. 1987;**4**(4):406-425. Epub 1987/07/01. DOI: 10.1093/oxfordjournals.molbev.a040454

[18] Harkins KM, Schwartz RS, Cartwright RA, Stone AC. Phylogenomic reconstruction supports supercontinent origins for Leishmania. Infection, Genetics and Evolution. 2016;**38**:101-109. Epub 2015/12/29. DOI: 10.1016/j.meegid.2015.11.030

[19] Benson DA, Cavanaugh M, Clark K, Karsch-Mizrachi I, Lipman DJ, Ostell J, et al. GenBank. Nucleic Acids Research. 2013;**41**(Database issue):D36-D42. Epub 2012/11/30. DOI: 10.1093/nar/gks1195

[20] Aslett M, Aurrecoechea C, Berriman M, Brestelli J, Brunk BP, Carrington M, et al. TriTrypDB: A functional genomic resource for the trypanosomatidae. Nucleic Acids Research. 2010;**38**(Database issue):D457-D462. Epub 2009/10/22. DOI: 10.1093/nar/gkp851

[21] Cabral AD, Garcia FB, Suzuki RB, Gois Filho TL, da Costa RT, Vasconcelos LMP, et al. Dataset on recombinant expression of an ancient chitinase gene from different species of Leishmania parasites in bacteria and in Spodoptera frugiperda cells using baculovirus. Data in Brief. 2020;**32**:106259. Epub 20200902. DOI: 10.1016/j.dib.2020.106259

[22] Schonian G. Genetics and evolution of Leishmania parasites. Infection, Genetics and Evolution. 2017;**50**:93-94. Epub 2017/04/01. DOI: 10.1016/j.meegid.2017.03.016

[23] van Eys GJ, Schoone GJ, Kroon NC, Ebeling SB. Sequence analysis of small subunit ribosomal RNA genes and its use for detection and identification of Leishmania parasites. Molecular and Biochemical Parasitology. 1992;**51**(1):133-142 Epub 1992/03/01

[24] Schonian G, Nasereddin A, Dinse N, Schweynoch C, Schallig HD, Presber W, et al. PCR diagnosis and characterization of Leishmania in local and imported clinical samples. Diagnostic Microbiology and Infectious Disease. 2003;**47**(1):349-358. DOI: 10.1016/s0732-8893(03)00093-2

[25] Cortes S, Rolao N, Ramada J, Campino L. PCR as a rapid and sensitive tool in the diagnosis of human and canine leishmaniasis using Leishmania donovani s.l.-specific kinetoplastid primers. Transactions of the Royal Society of Tropical Medicine and Hygiene. 2004;**98**(1):12-17 Epub 2004/01/02

[26] Paiva BR, Passos LN, Falqueto A, Malafronte Rdos S, Andrade HF Jr. Single

step polymerase chain reaction (PCR) for the diagnosis of the Leishmania (Viannia) subgenus. Revista do Instituto de Medicina Tropical de São Paulo. 2004;**46**(6):335-338. Epub 2005/01/18. DOI: S0036-46652004000600007

[27] da Silva LA, de Sousa CS, da Graca GC, Porrozzi R, Cupolillo E. Sequence analysis and PCR-RFLP profiling of the hsp70 gene as a valuable tool for identifying Leishmania species associated with human leishmaniasis in Brazil. Infection, Genetics and Evolution. 2010;**10**(1):77-83. Epub 2009/11/17. DOI: 10.1016/j.meegid.2009.11.001

[28] Degrave W, Fernandes O, Campbell D, Bozza M, Lopes U. Use of molecular probes and PCR for detection and typing of Leishmania - a mini-review. Memórias do Instituto Oswaldo Cruz. 1994;**89**:463-469

[29] Suzuki RB, Cabral AD, Martins LP, Speranca MA. A highly sensitive Leishmania infantum chagasi isolation method from bone marrow and peripheral blood of adults and children. Journal of Infection in Developing Countries. 2016;**10**(11):1275-1277. Epub 2016/11/26. DOI: 10.3855/jidc.8022

[30] Hijjawi N, Kanani KA, Rasheed M, Atoum M, Abdel-Dayem M, Irhimeh MR. Molecular diagnosis and identification of leishmania species in Jordan from saved dry samples. BioMed Research International. 2016;**2016**:6871739. Epub 2016/07/13. DOI: 10.1155/2016/6871739

[31] Razek-Desouky A, Specht CA, Soong L, Vinetz JM. Leishmania donovani: Expression and characterization of Escherichia coli-expressed recombinant chitinase LdCHT1. Experimental Parasitology. 2001;**99**(4):220-225. Epub 2002/03/13. DOI: 10.1006/expr.2001.4665

[32] Button LL, Wilson G, Astell CR, McMaster WR. Recombinant Leishmania surface glycoprotein GP63 is secreted in the baculovirus expression system as a latent metalloproteinase. Gene. 1993;**134**(1):75-81 Epub 1993/11/30

[33] Saibil H. Chaperone machines for protein folding, unfolding and disaggregation. Nature Reviews. Molecular Cell Biology. 2013;**14**(10):630-642. Epub 2013/09/13. DOI: 10.1038/nrm3658

[34] Clare DK, Saibil HR. ATP-driven molecular chaperone machines. Biopolymers. 2013;**99**(11):846-859. Epub 2013/07/24. DOI: 10.1002/bip.22361

[35] Requena JM, Montalvo AM, Fraga J. Molecular chaperones of Leishmania: Central players in many stress-related and -unrelated physiological processes. BioMed Research International. 2015;**2015**:301326. Epub 2015/07/15. DOI: 10.1155/2015/301326

[36] Mukherjee D, Patra H, Laskar A, Dasgupta A, Maiti NC, Datta AK. Cyclophilin-mediated reactivation pathway of inactive adenosine kinase aggregates. Archives of Biochemistry and Biophysics. 2013;**537**(1):82-90. Epub 2013/07/09. DOI: 10.1016/j.abb.2013.06.018

[37] Wang C, Wang H, Zhang D, Luo W, Liu R, Xu D, et al. Phosphorylation of ULK1 affects autophagosome fusion and links chaperone-mediated autophagy to macroautophagy. Nature Communications. 2018;**9**(1):3492. Epub 2018/08/30. DOI: 10.1038/s41467-018-05449-1